Seacoast Life

SEACOAST LIFE

An Ecological Guide to Natural Seashore Communities in North Carolina

by Judith M. Spitsbergen

Illustrations by Jean Hoxie,
Mary Ann Nelson, Will Thomson

The University of
North Carolina Press · Chapel Hill

Published for the
North Carolina State Museum of Natural History, Raleigh,
in association with the Hampton Mariners Museum, Beaufort

Reprinted 1983 by the University of North Carolina Press by special arrangement with the North Carolina State Museum of Natural History

94 93 92 9 8 7 6

This work was partially sponsored by the Office of Sea Grant, NOAA, U.S. Department of Commerce, under Grant No. 04-6-158-44054, and the North Carolina Department of Administration. The U.S. Government is authorized to produce and distribute reprints for governmental purposes notwithstanding any copyright that might appear hereon.

The Hampton Mariners Museum, Beaufort, North Carolina, is an extension of the North Carolina State Museum of Natural History, a division of the North Carolina Department of Agriculture, James A. Graham, Commissioner

Book design: Lynn Repasky
Cover design: Tim Dove

LC 83-80687
ISBN 0-8078-4109-9

CONTENTS

ACKNOWLEDGMENTS

The author wishes to recognize the following persons who filled their own valuable niches in the completion of this book:

For his influence, thoughts, and words which are reflected throughout the book, Will Hon, Director of Educational Programs, UGA Marine Resources Extension Center, Skidaway Island, Georgia.

For their review of scientific content, Bill Kirby-Smith, Duke University Marine Laboratory, Beaufort, North Carolina, and Gordon Thayer, National Marine Fisheries Service, Beaufort.

For their art and illustrations, Jean Hoxie, Beaufort, Mary Ann Nelson, Duke University Marine Laboratory, Beaufort, and Will Thomson, North Carolina State Museum of Natural History, Raleigh.

To Sea Grant, B.J. Copeland, Director, for its financial assistance with illustrations.

For their fine editing, Bill Nicholson, National Marine Fisheries Service, Beaufort, John E. Cooper and Alexa Williams, North Carolina State Museum of Natural History.

For their enthusiastic sponsorship of the book, John B. Funderburg, Director, North Carolina State Museum of Natural History, and Charles McNeill, Curator, Hampton Mariners Museum, Beaufort.

And for his consistent moral support, Norman Anderson, Department of Math and Science Education, North Carolina State University, Raleigh.

PREFACE

Seacoast Life, designed primarily for ecological field study, will be helpful to teachers, students, and anyone who likes to explore the seacoast. This guide also can be used as a classroom text and as a resource to prepare for field trips to the coast.

The primary focus of the book is on the ecology of seashore and estuarine communities rather than on the identification of individual plants or animals. Since it is impossible to include all species in a community, those selected for illustration are the ones most likely to be encountered in the field. Some species, though seldom seen, are included because they are unique; others visible only with a microscope are included because they perform vital roles in the community.

Often the particular genus illustrated represents several similar genera. The abbreviation "sp." means that the precise species is not indicated, and "spp." means that several species of the genus occur. For the sake of brevity, species commonly found in more than one type of community usually are not repeated. Discussions of plant and animal species include common names, when available, as well as scientific names. The organisms, habitats, and communities included are typical of the mid-Atlantic coast.

It was not possible to depict food webs and energy flow patterns in each community. Since these intricate subjects are, for the most part, still being studied and debated among ecologists, they are beyond the scope of this guide.

The section on seashore ecology is basic to understanding other sections in the guide. A bibliography is included for those interested in more detailed study of community ecology. A glossary defines terms that may be unfamiliar.

The author is an educator who enjoys sharing the adventure of learning about North Carolina's marine environments. She is curator of education at the Hampton Mariners Museum, Beaufort, North Carolina, a division of the North Carolina State Museum of Natural History, and frequently takes groups on field trips for the same ecological studies presented in this guide.

Seacoast Life

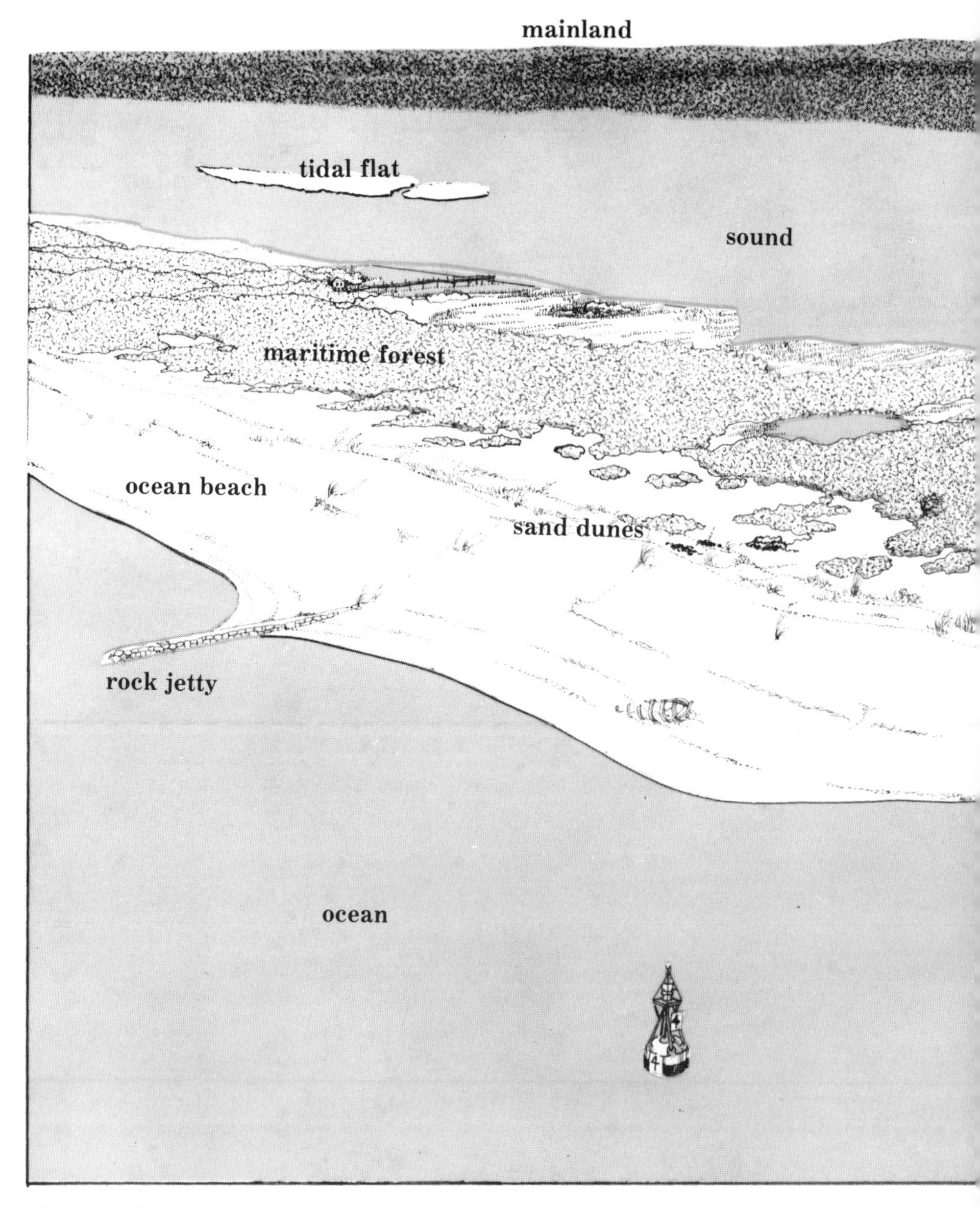

Figure 1. The coastal landscape is the setting for shoreline communities.

Seacoast Ecology

Looking at the landscape

The coast of North Carolina is fringed by a string of barrier islands separated from the mainland by shallow sounds. Narrow ribbonlike patterns of ocean beaches, sand dunes, maritime forests, salt marshes, and tidal flats extend for hundreds of miles along the coast (Figure 1). Man-made jetties, seawalls, docks, and bridges dot the shoreline, interrupting the natural scenery. While all these diverse coastal settings blend together in the landscape, each type is a distinctly separate shoreline habitat, quite different from the others. In each develops a natural community of plant and animal life unique to that type of habitat.

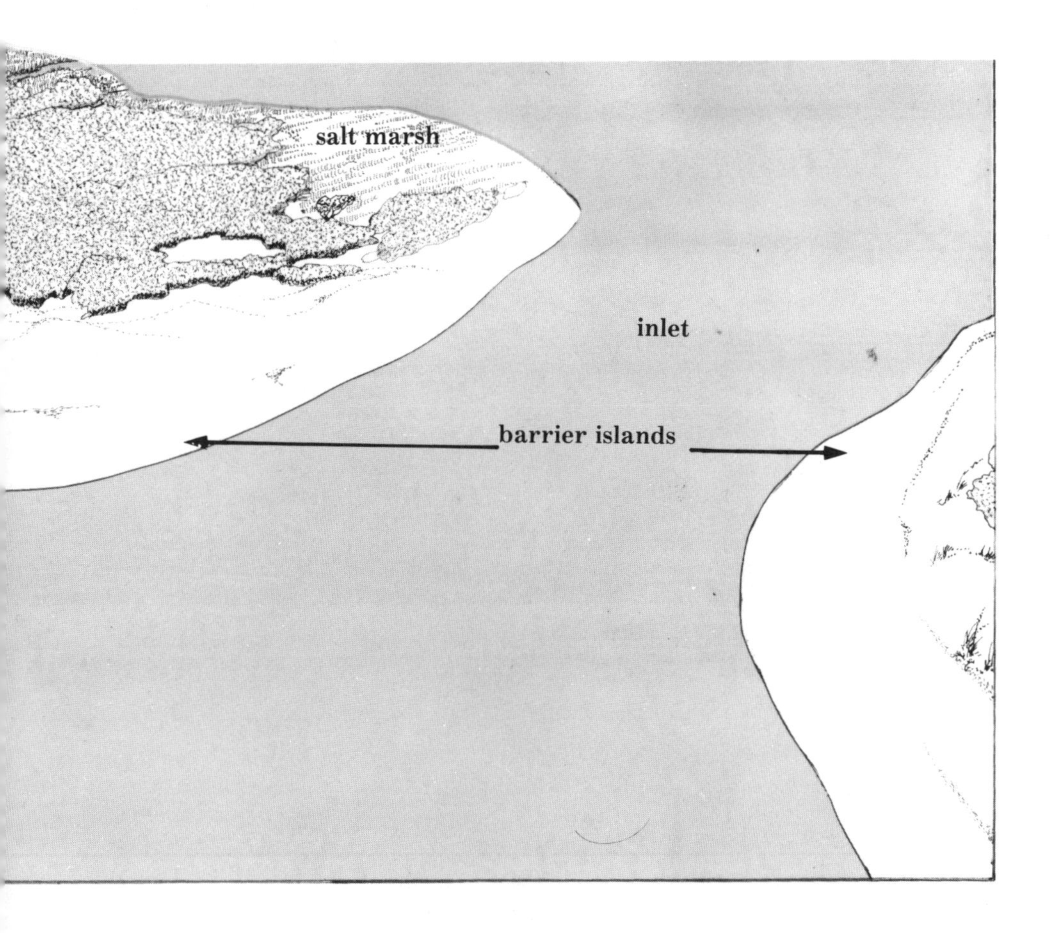

Coastal habitats

Ecology is the study of relationships among plants and animals and their environment, which includes all living and nonliving things that influence organisms. Within any major environment are many different habitats where plants and animals live. A habitat is a smaller unit of the total environment and consists of the area and conditions immediately surrounding an organism. To survive in its habitat an organism must be adapted to the conditions existing there. In shoreline habitats the chief environmental factors include climate, water level, salinity, water and air temperatures, and substrate.

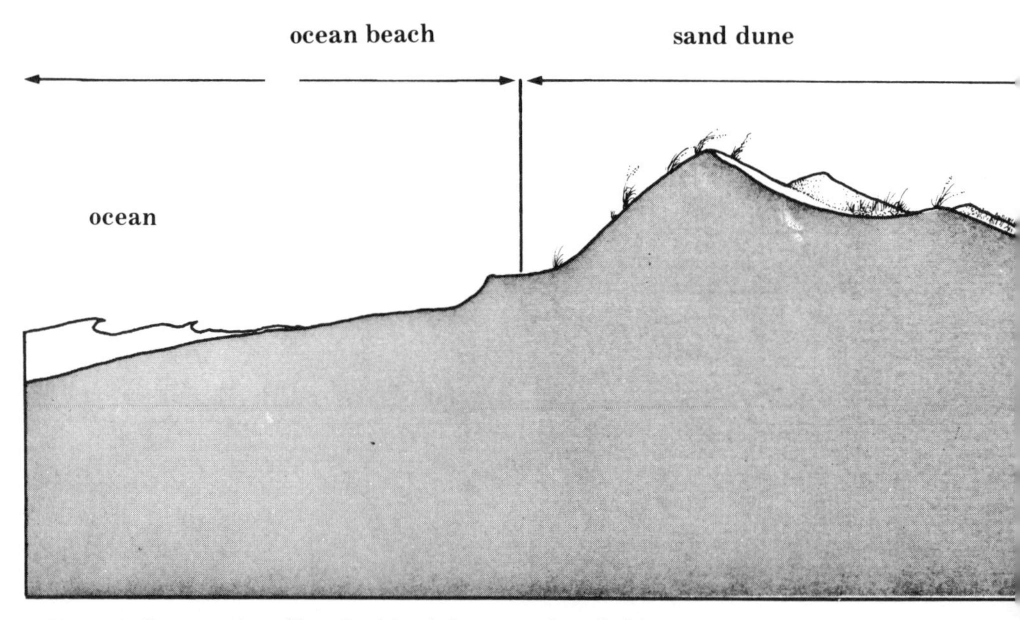

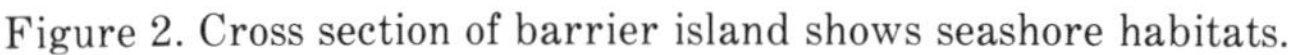
Figure 2. Cross section of barrier island shows seashore habitats.

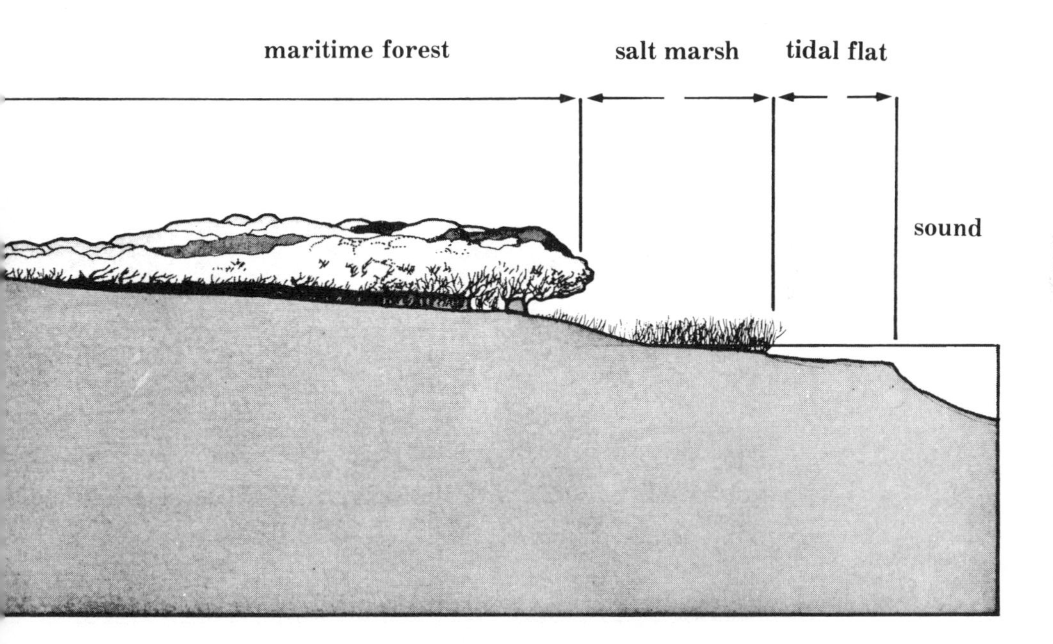

The seashore habitats of North Carolina's coast are subject to many of the same environmental conditions. The habitats include sandy beaches fronting the ocean, sand dunes and maritime forests generally existing on the barrier islands, salt marshes and tidal flats bordering the coastal sounds, and man-made solid structures such as jetties and pilings projecting from the shoreline (Figure 2). These different habitats may occur adjacent to one another and share the same general climate. However, variations in water level, temperature, salinity, and substrate can have a profound influence on plants and animals living in seashore habitats.

Climate

Coastal North Carolina has a moderate climate. South of Cape Hatteras temperatures are kept relatively warm the year around by Gulf Stream waters flowing up from the south. North of Cape Hatteras temperatures are slightly cooler because the Virginia Current, an offshoot of the Labrador Current, brings in colder waters from the north (Figure 3). As a result, somewhat different plant and animal species may be found north and south of the cape.

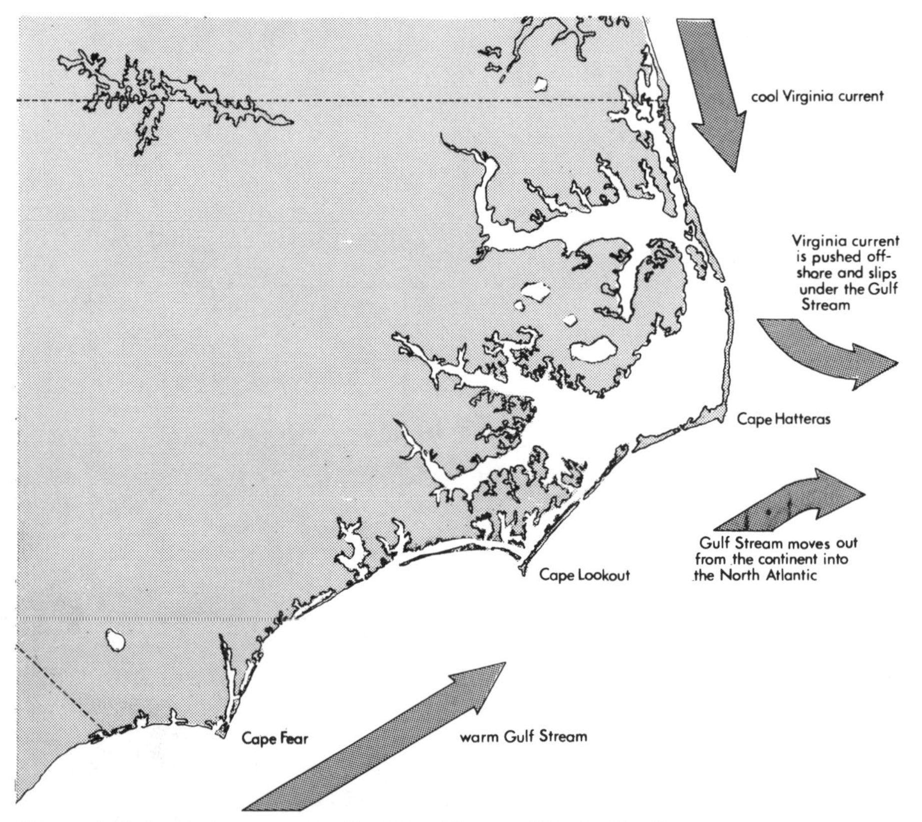

Figure 3. Major water currents affect the climate of North Carolina.

Water level

Water level exerts a strong influence on shoreline habitats. As the tides rise and fall twice each 24 hour period, the water level alternately exposes and submerges the shoreline. On the North Carolina coast the tidal range is between 0.6 and 1.3 meters, depending on the general location. Seashore habitats can be divided into zones based on the tides (Figure 4). The subtidal zone is always covered by water, and organisms living in this zone are unable to survive when exposed to air. Between the low and high tide lines lies the intertidal zone, alternately covered by water and exposed to air. Organisms living in this zone must be adapted to drastically changing conditions during the tidal cycle. Tidal water does not reach above the high tide line into the supratidal zone, so organisms here are essentially terrestrial rather than aquatic. Elevation and slope of the land determine how far the zones extend in a habitat.

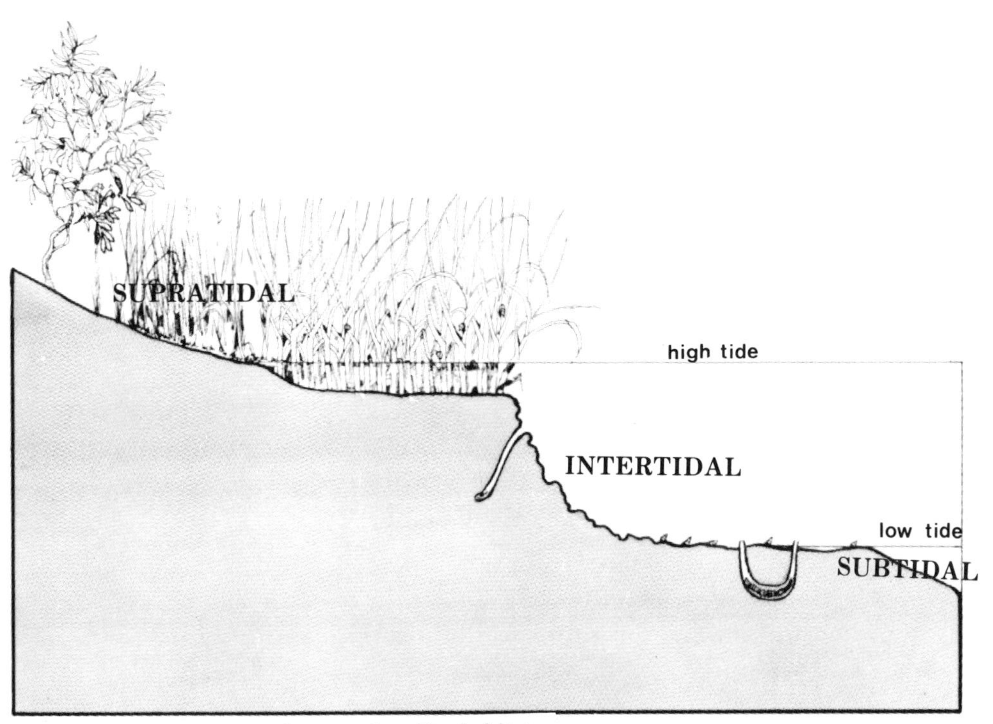

Figure 4. Tidal range influences shoreline habitats.

Salinity

Because most organisms cannot tolerate changes in salinity, their distribution is strongly controlled by the amount of salt in the water. A few species can withstand a wide range of salinity and can survive in both sound and ocean habitats. The average salinity of ocean water is 35 parts dissolved salts per thousand parts of water, or 35 0/00 (Figure 5). Rising tides flush large volumes of ocean water through inlets into sounds. Fresh river water also empties into the sounds, and water near a river mouth may be quite brackish. Areas, such as these sounds, where fresh and salt water mix are called estuaries. The water in estuaries is less salty, generally 15 to 25 0/00.

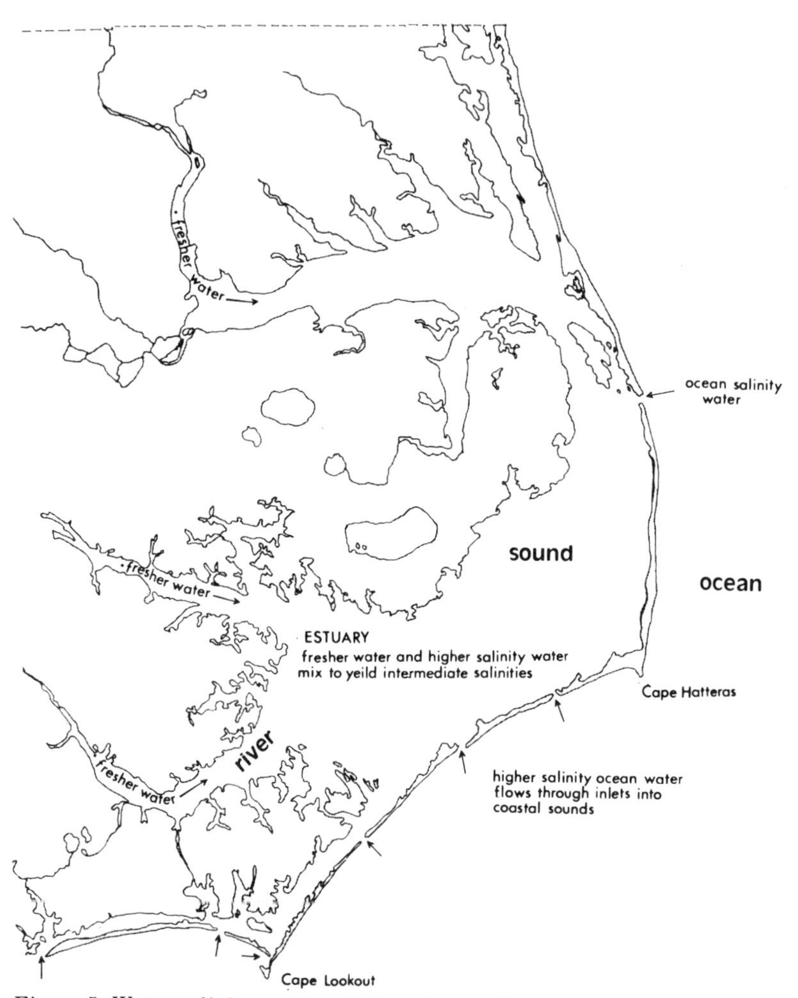

Figure 5. Water salinity varies with location on the coast.

Temperature

Shoreline habitats may be subjected to wide extremes and rapid changes in water temperature during the tidal cycle. Water temperatures do not change as rapidly as air temperatures, and ocean water temperatures remain fairly constant during a given season. Surface water, however, may be quite a bit warmer than deeper water on a hot summer day, or cooler during winter. In shallow coastal sounds the water may warm or cool much more rapidly than in the ocean. Generally, incoming tides bring cool ocean water to sounds, and rivers bring warmer or cooler water into estuaries, depending on the season. Organisms living in shoreline habitats must be adapted to these changes and variations.

Substrate

Substrate, the bottom on or in which an organism may live, influences marine life. Most plant and animal species prefer a particular type of substrate and live only in habitats where that type is available. Some species require sand or mud; others require more solid materials such as rocks or shell remains. The type of substrate characteristic of shoreline habitats varies with water turbulence (Figure 6.) Coarse sediments settle to the bottom and form the substrate on an ocean beach where water is most turbulent. Mud, silt, and finer-grained sands settle on the bottom of sounds where less turbulent water flows. Large expanses of rocky substrate do not occur naturally on the North Carolina coast, and competition is therefore great among species that must live attached to solid surfaces.

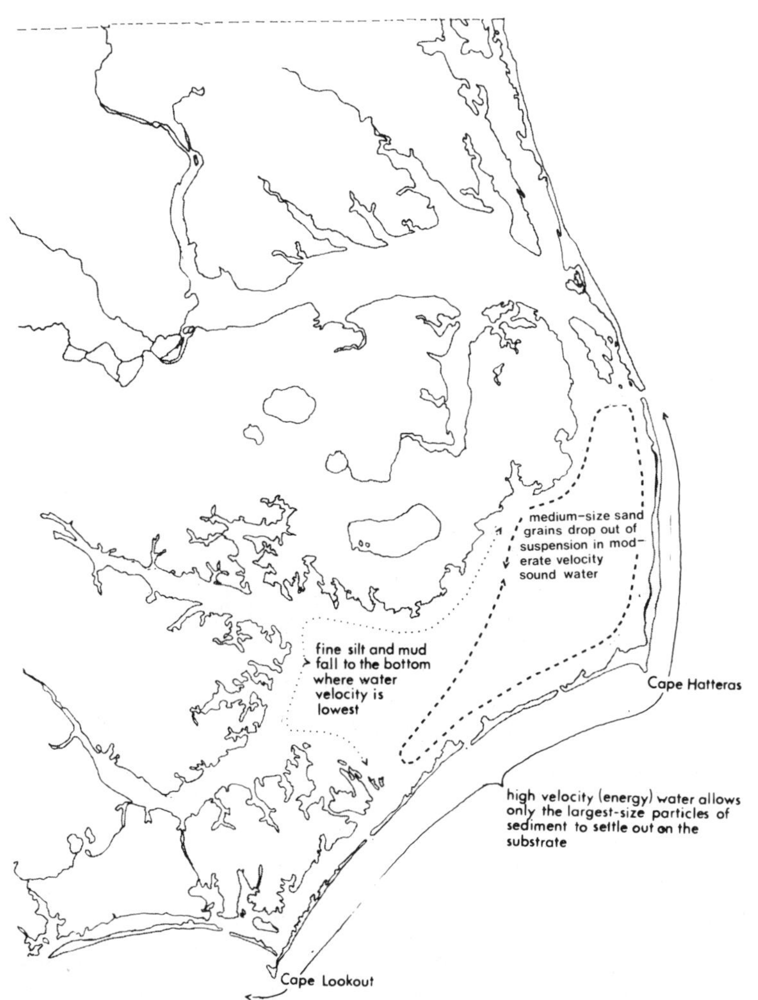

Figure 6. Water velocity determines size of sedimentary particles.

Habitat and community

To survive in any habitat, plants and animals must develop methods of living with and among other species; thus, many types of interactions occur. The relationships among species living together in a habitat result in the development there of an ecological community. For example, a sand dune community forms in the habitat of beach dunes; a salt marsh community forms in the marshy shoreline of sounds.

An ecological community is like a "unit of life", with plant and animal components acting together as an interdependent group. The community concept helps explain how and possibly why a variety of plants and animals are found living together in an orderly manner instead of functioning haphazardly. All communities form and operate on the same basic ecological principles, regardless of the habitat in which they form.

The ecological community concept provides a means of looking at nature in a more complete, meaningful way. In studying a community the primary emphasis is not on learning the names of plants or animals, although identifying each organism is vital to understanding a community. The concept goes further to ask how one organism in a habitat relates to other organisms, and ultimately how all are interrelated. The ecological community is a unit of life and provides, on a small scale, a framework for looking at the greater principles of nature. Although the structure and functioning of any natural community cannot be completely explained, the community concept has become one of the most fruitful tools of ecology.

Food and energy in the community

An ecological community develops primarily around feeding patterns of organisms living together in a habitat. Like a complex machine, a community consists of many parts, each requiring energy to keep it "working" and sustain its life. Energy is passed through a community by means of feeding relationships. Each species has a niche — a role it plays — that fits into the cycle of food and the flow of energy through the community. Because this cycle is complex, scientists call it a "food web."

A community food web has three basic components — producers, consumers, and decomposers (Figure 7). The relative number and variety of plant and animal species in each component are important in the community.

Producers

Producers are green plants that trap and store energy from sunlight, the original energy source on which the whole community depends. Through photosynthesis, plants "produce" organic material which they make from carbon dioxide, water, and inorganic nutrients available in the environment. During the same process these producers convert radiant energy, storing it as "fixed" energy in the organic matter of their body tissues. The source and abundance of inorganic nutrients, the amount of energy stored in the tissues of producers, and the number and type of producers all play a vital part in the community. Producers in seashore communities consist of microscopic and large algae and rooted plants that have adapted to living in salty water. Most producers in the sand dune community and in the supratidal zone of other communities are rooted terrestrial plants.

Consumers

Consumers are animals which must eat plants (herbivores), or eat other animals (carnivores), or eat both (omnivores) to obtain nutrients and energy. Consumers in seashore communities range in size from microscopic to large animals such as birds, fishes, and others even larger. In a community there may be four, or rarely five, feeding (trophic) levels of consumers, usually made up of increasingly larger animals feeding on prey smaller than themselves. In the consumer-eats-prey relationships, energy and nutrients contained in the tissues of organisms at one feeding level are transferred to organisms at the next feeding level. Some of the transferred energy and nutrients are stored as organic food energy in the body tissues of the consumers. However, the organisms at each feeding level use most of the energy they consume to carry out their own life processes. Therefore, the number of feeding levels is limited by the progressive loss or use of energy at each "step" in the food web. At each feeding level the source and abundance of food is very important in the community. The existence of animals with different feeding methods limits excessive competition for the same resource. Figure 8 shows some of the basic feeding methods common in marine communities.

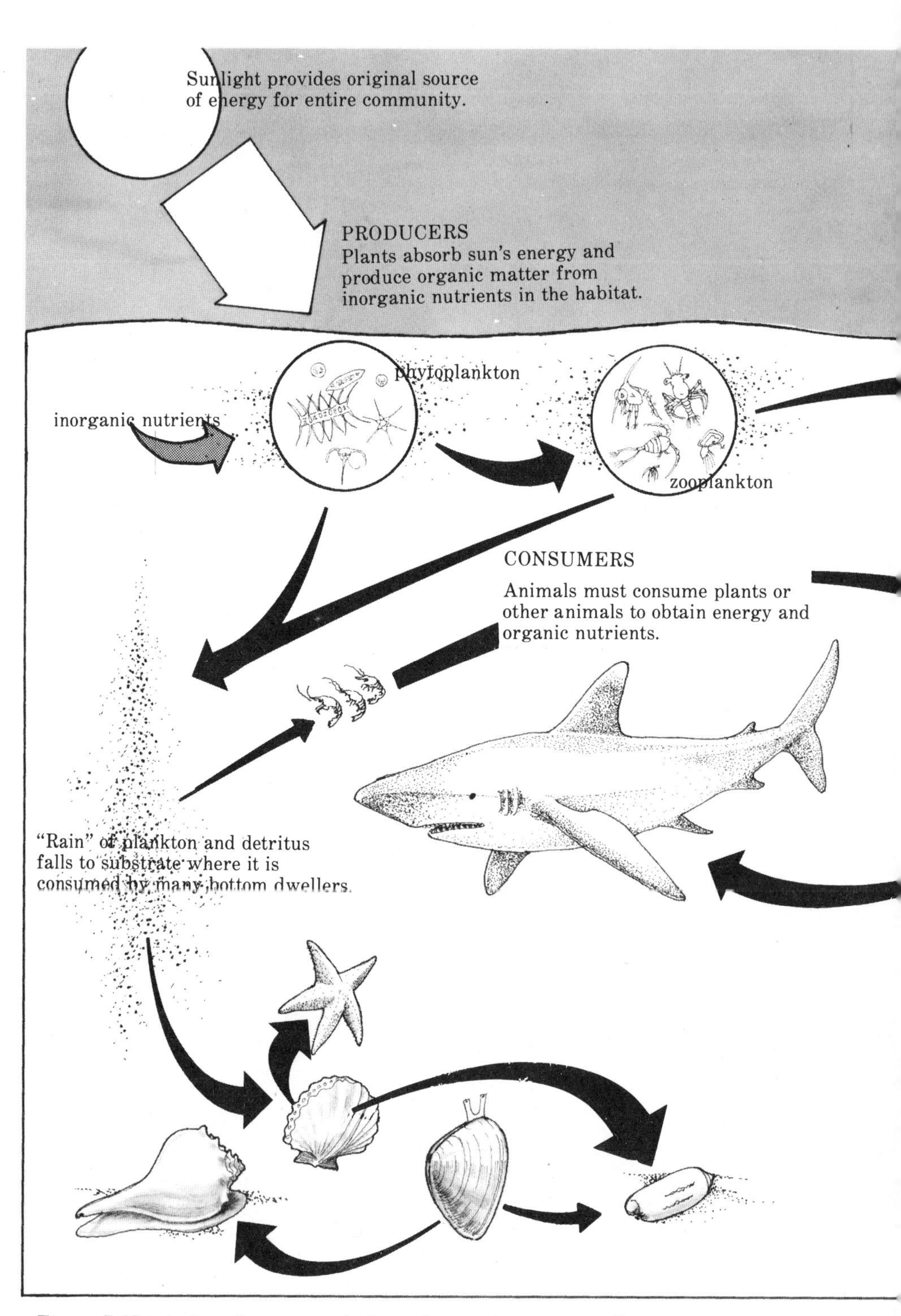

Figure 7. Nutrients and energy cycle through a seashore community.

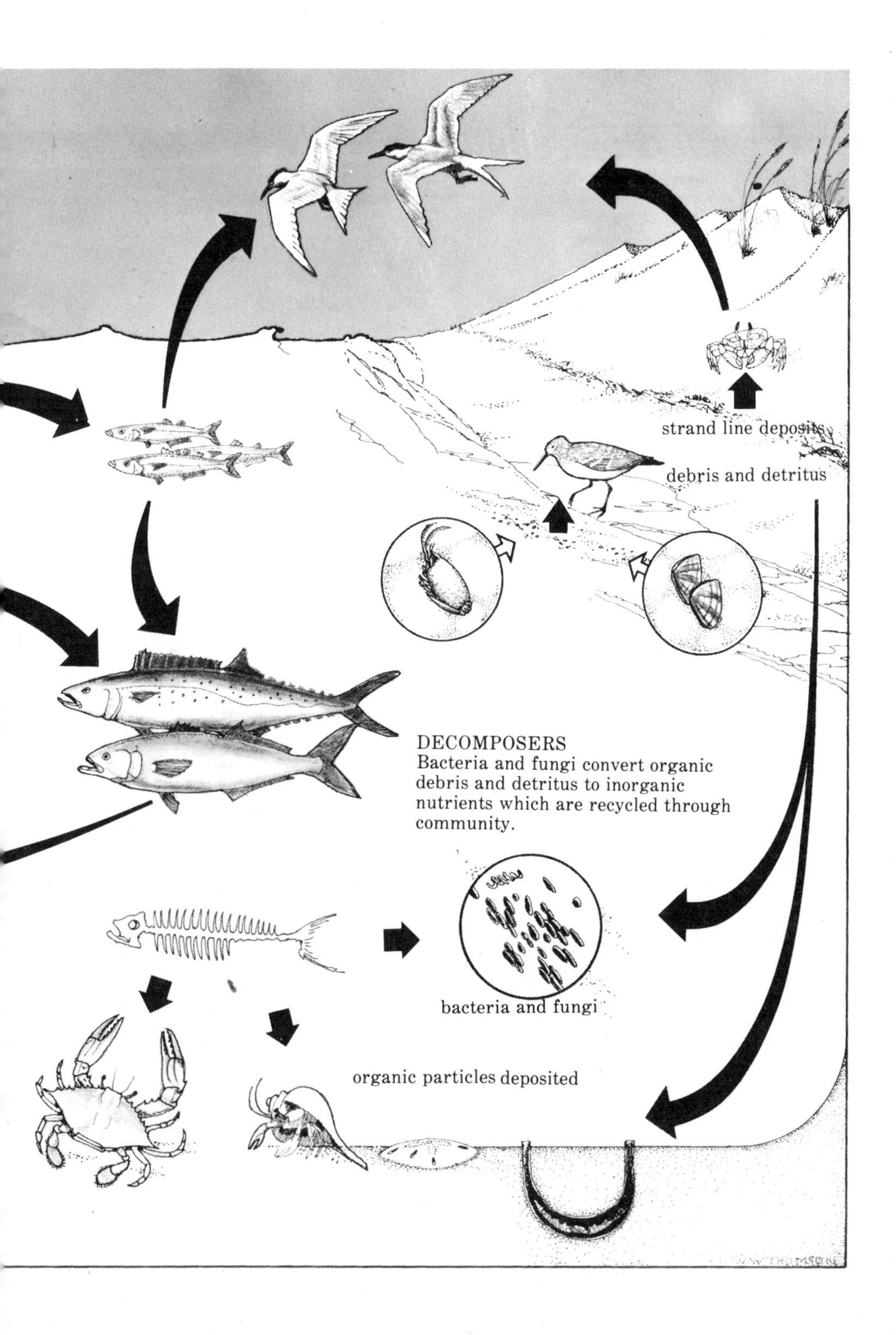
strand line deposits
debris and detritus
DECOMPOSERS
Bacteria and fungi convert organic debris and detritus to inorganic nutrients which are recycled through community.
bacteria and fungi
organic particles deposited

Food capturers

These predators, such as the bluefish, actively chase or ensnare prey. Most food capturers move rapidly and have well-developed vision. Some, such as the squid, have tentacles that ensnare food.

Food finders

Some animals, such as the limpet, slowly graze over surfaces, scraping away and eating small attached organisms. Other more "casual" predators, such as the hermit crab, move slowly to find and/or scavenge their food.

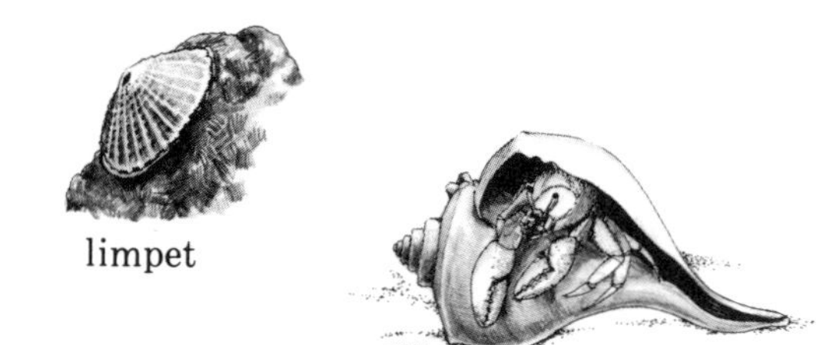

Deposit feeders

These animals feed on materials that have settled to the bottom. Some, including the sea cucumber, have siphon tubes and function like biological vacuum cleaners to suck up deposited material. Others, called sediment ingesters, eat quantities of sediments, extract organic matter, and then defecate unused sediments. The lugworm is an example.

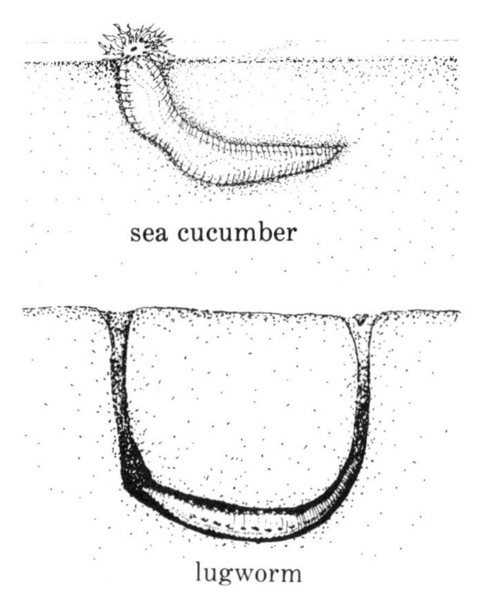

Figure 8. These feeding methods are common among animals in marine communities.

Suspension feeders

These animals eat suspended particles in water, "screening" large quantities of water to extract food, usually plankton. The following types of feeders are all suspension feeders.

Arthropod type:

Often called "filter feeders", these organisms sweep and strain water with feather-like appendages or with special rakers that filter water passing over the gills.

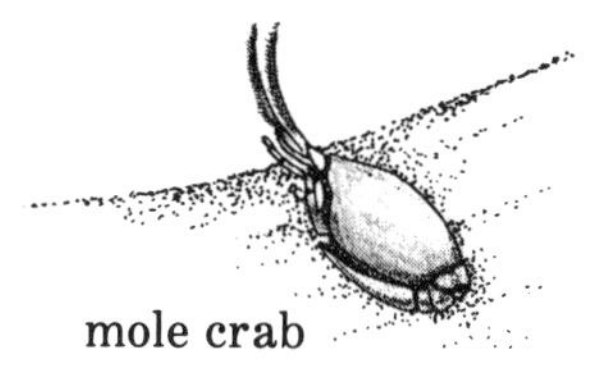
mole crab

Mollusk type:

These animals pump streams of water through their bodies, removing suspended food particles with a remarkable mucous-trapping device for straining water. Incoming water passes over the sticky, mucous-covered gill surfaces, where particles become trapped. Food and mucous then are carried by cilia to the mouth and stomach.

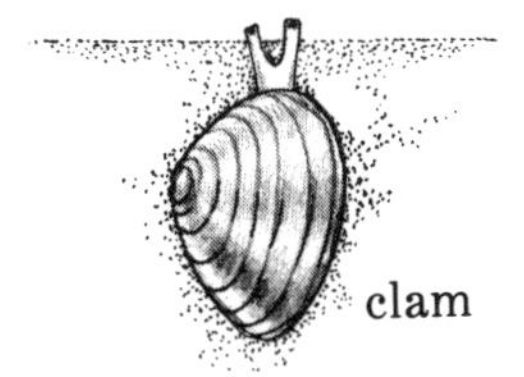
clam

Tentacle type:

These animals wave their tentacles in water to catch suspended organisms and food particles. Mucous glands, and in some cases stinging cells, on the tentacles aid in the capture.

sea anemone

Decomposers

Decomposers, mostly bacteria and fungi, accomplish the final step in any food web. This is the breakdown of plant and animal organic tissue and the return of inorganic materials to the environment. Other natural environmental processes also play a part. As plants and animals die, their remains first become decaying debris that floats about in the water or settles to the substrate where it quickly attracts scavengers and decomposers. Scavengers consume much of the organic debris, particularly animal tissue. Uneaten debris is further broken down by the action of waves, wind, heat, and water into very small particles of organic matter called detritus. Bits of detritus, along with the bacteria and fungi embedded in it, float in the water or settle on the substrate and provide an important energy source for many organisms which feed on it. Through the action of decomposers and of these processes, organic debris and detritus ultimately are decomposed into inorganic nutrients which become available to plants for photosynthesis. Thus, the cycling of nutrients and energy through the community begins again.

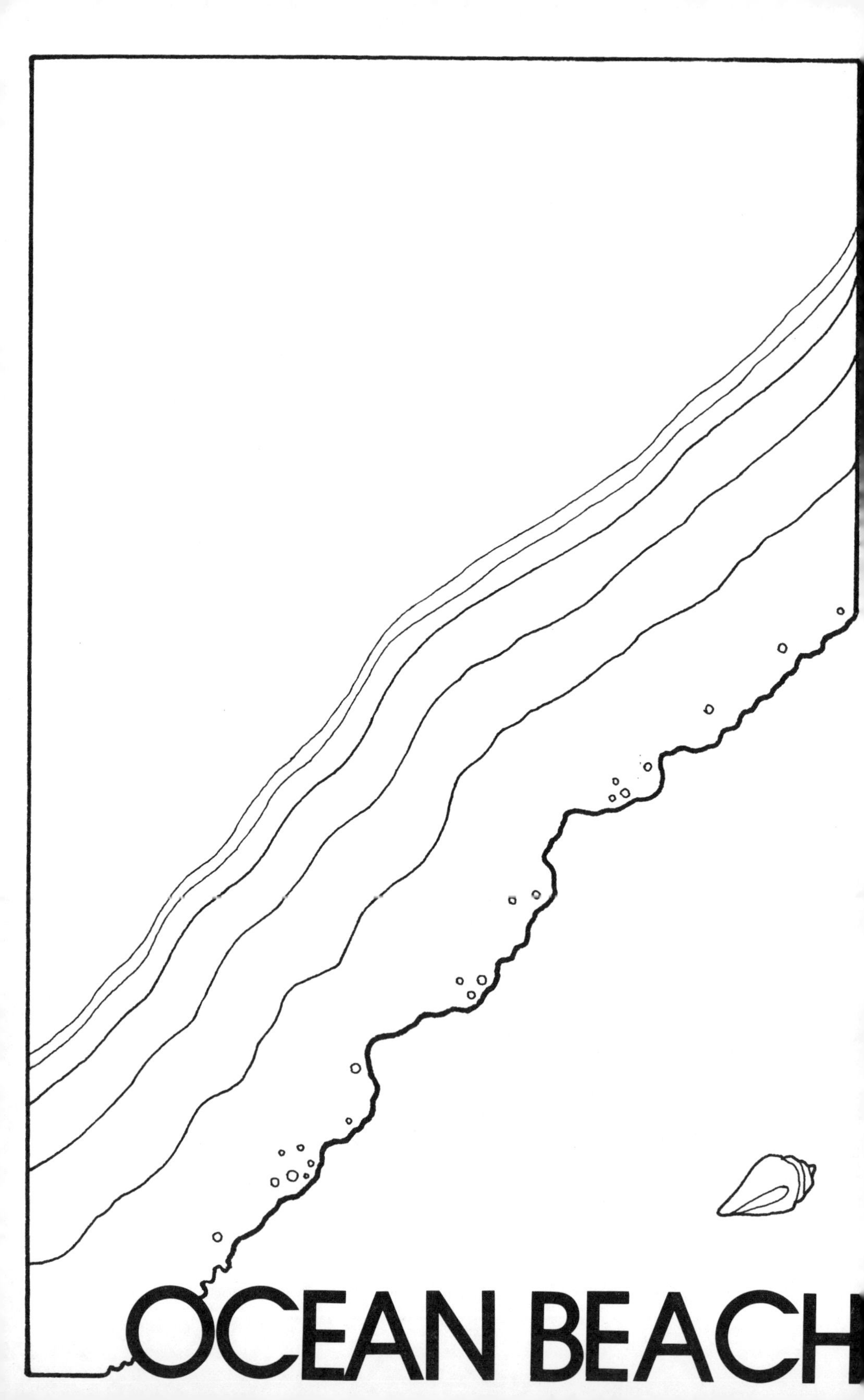
OCEAN BEACH

OCEAN BEACH HABITAT

The habitat

Ocean beach habitats on barrier islands have three distinct zones: the subtidal zone of crashing breakers with swirling sand and shallow water; the intertidal zone, alternately covered and exposed by tidal waters; and the dry, sandy supratidal zone extending to the base of the sand dunes. The appearance of the sand beach, directly exposed to strong winds and ocean waves, may change from day to day. Because of the intense physical forces, plants and animals in this habitat are highly specialized and only a small variety of species occurs here. Those adapted to the habitat, however, occur in tremendous numbers.

Special features

—The powerful surf pounds the subtidal and intertidal substrate.
—Wave action sweeps larger shell particles and sand across the substrate surface, sometimes causing abrasions or burying plants and animals.
—Shifting sand in the surf prevents attaching organisms from obtaining a hold.
—Water temperatures throughout the tidal cycle are fairly constant from day to day but may vary considerably between seasons. Water surface temperatures are about the same as air temperatures.
—Water salinity shows little variation and is near average ocean salinity. Estuarine water flowing out on the ebbing tide may reduce water salinities near inlets.
—Oxygen is plentiful in the water and sediments.
—Much of the intertidal substrate is inundated every few hours and remains fairly moist.
—The intertidal and supratidal zones offer little or no protective cover to organisms.
—Surface temperatures of the supratidal sand are excessively high during the hot summer months, but temperatures below the surface are moderate.
—Rainwater striking the supratidal sand quickly percolates into the surface and below ground. The natural water table may be 2.0 meters or more below the dry surface sand where evaporation is rapid. Below the surface, where the sun does not evaporate moisture, humidity is moderate.
—Debris and detritus, deposited at the high tide line by spring and neap tides, often accumulate in two evident strand lines.
—Strong winds characteristic of the ocean beach often spread a cutting barrage of fine sand particles. There is no cover protection to modify the harsh effects of these winds.

Adaptations of plants and animals

—Abundant microscopic animals exist in a unique micro-habitat among the damp sand particles of the subtidal and intertidal beaches.
—Many animals in the subtidal zone are rapid burrowers that dig into the substrate to avoid shifting surface particles. Others are efficient swimmers, well-adapted to turbulent water. Many swimmers also burrow temporarily into the bottom to insure a "sand cover" protection.
—Many subtidal animals are suspension feeders, particularly crabs and clams and their relatives. Their mouth and body parts are adapted for extracting plankton from sea water.
—Few animals which build permanent tubes or burrows are found on the subtidal and intertidal beach, since shifting sediments quickly clog and cover such dwellings.
—Few sediment ingestors live in any zone of the beach because nutrient detritus does not accumulate on the substrate surface.
—Intertidal animals move easily with the shifting sand and rushing waves. After being swept out of the sand by advancing or receding waves, they rapidly burrow in again.
—Birds fly or run quickly ahead of the advancing intertidal surf, feeding on abundant small forms of life.
—Supratidal organisms dig burrows in the sand to avoid extreme temperatures and arid conditions on the surface.
—Most supratidal animals are nocturnal, feeding and foraging when darkness provides more moderate temperatures and some protection from predators.

The community

In the subtidal zone bountiful microscopic plants called phytoplankton are the main producers, nourished by plentiful inorganic nutrients. Microscopic animals called zooplankton occur abundantly in the surf, feeding on phytoplankton and other zooplankton. Suspended plankton provide food for the many filter feeding animals in the community. By contrast, there are few deposit feeders since water velocity prevents detritus from settling on the sandy bottom. Myriad microscopic life thrives in the sand and provides an important food source for other animals dwelling in the sandy bottom. Intertidal beach life is nourished largely by plankton, brought in by advancing tides. Since large plants or attached seaweeds generally require a more stable, solid substrate, they cannot live in the subtidal and intertidal zones where powerful ocean waves would uproot them. A noticeable lack of living plants characterizes the upper, dry supratidal beach. Most supratidal animals are scavengers that feed on accumulated debris and detritus along the strand line.

Typical organisms

Subtidal life

Plankton

Phytoplankton (phyto = plant, plankton = wanderer) are among the most important producers in the ocean. Great numbers of these free-floating microscopic plants are swept about by currents. The diatoms and dinoflagellates illustrated are but two of the seemingly infinite varieties of shapes and types. As the seasons change, different species become successively abundant, but for any given season the same species generally are present. The total volume of phytoplankton, however, tends to be greater in warm months. Plants such as these compose the "pastures of the sea;" eaten by tiny consumers, they form the basis of marine food webs.

Representative phytoplankton
(magnified about 100 X)

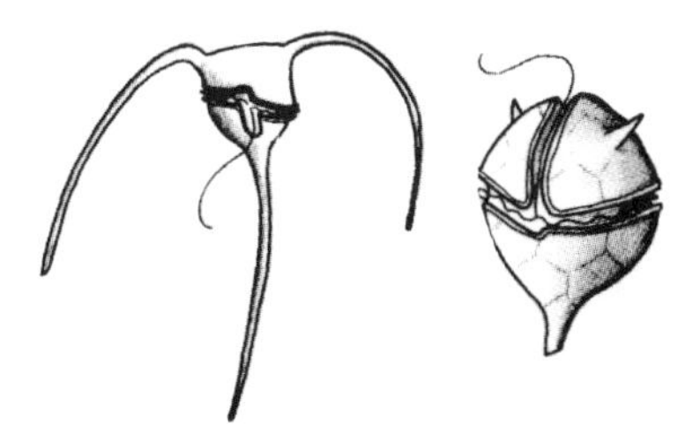

Dinoflagellates cause "red tides" when sudden blooms of certain species occur. Tiny whiplike flagellae in some species enable them to swirl about in the water. (Microscopic.)

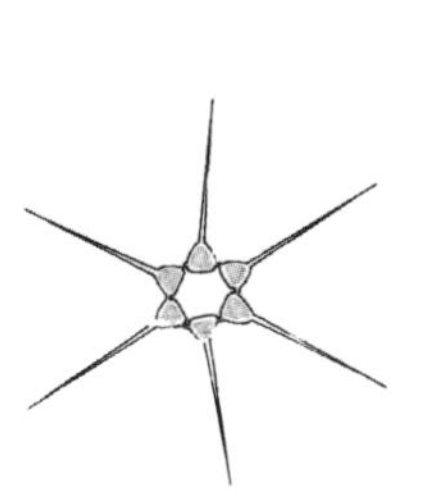

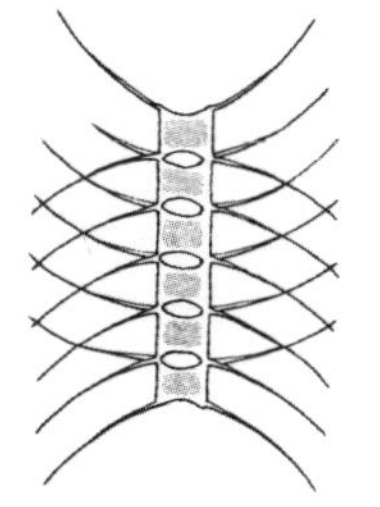

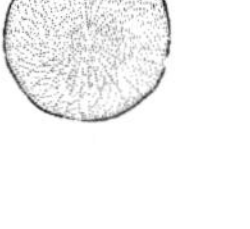

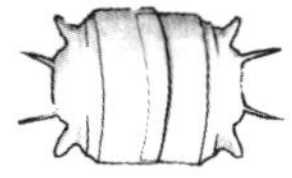

Diatoms have marvelously delicate "shells" of silica, displaying a tremendous variety of shapes. The transparent shell has an upper portion fitting into a lower one. (Microscopic.)

Zooplankton (zoo = animal, plankton = wanderer) are minute animals carried with ocean currents to the surf zone of the ocean beach. Occurring in great numbers, zooplankton feed on phytoplankton in the water or eat other zooplankton smaller than themselves. Such animals, therefore, form the second and third steps in marine food webs. Some zooplankton, like tiny copepods, isopods, and amphipods, remain small and planktonic all their lives. Others are planktonic only when young. The hatching or spawning of larval forms causes some seasonal variation in the types of zooplankton, and all types are an important source of food for other animals.

Representative zooplankton
(magnified about 10 X)

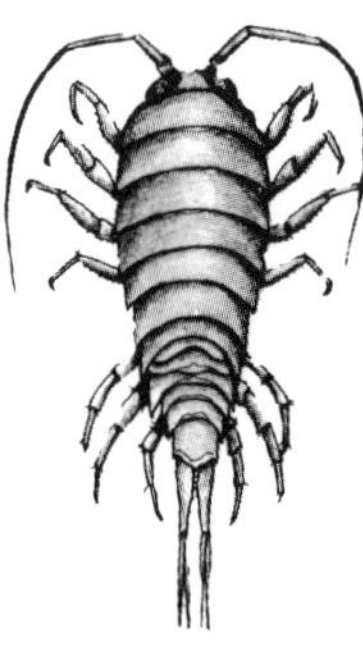

ISOPOD

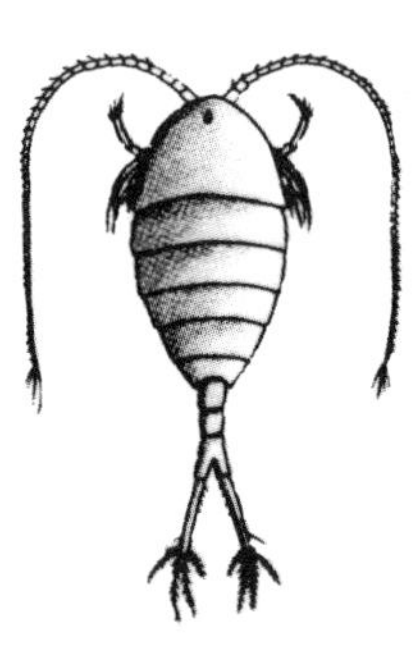

COPEPOD

AMPHIPOD

Larval and immature animals

Larval forms of marine animals are common among the plankton. Distinctive in appearance, they are quite different from the adults, and some stages have special names. Most of the forms shown below occur as ocean beach plankton at some time during the year.

LARVAL BARNACLE

nauplius stage

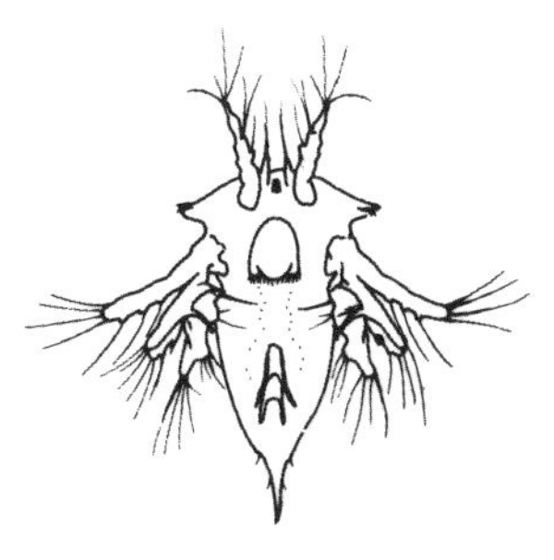

cypris stage

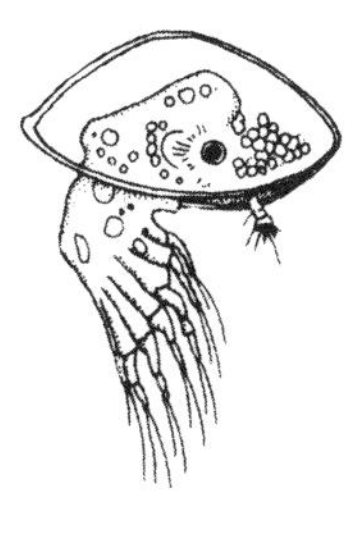

LARVAL MOLLUSC

veliger stage

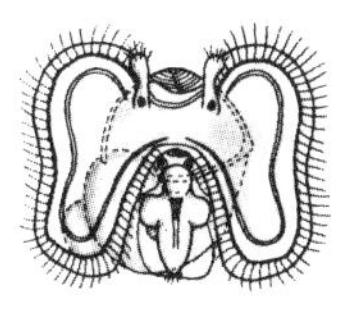

LARVAL WORM

trochophore stage

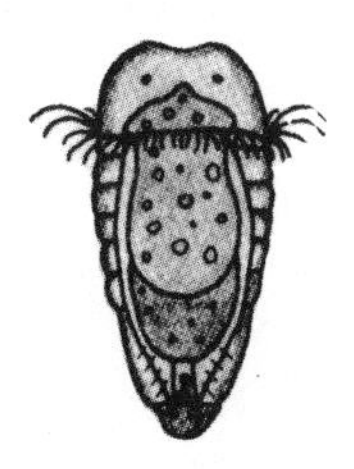

LARVAL HORSESHOE CRAB

eurypterid stage

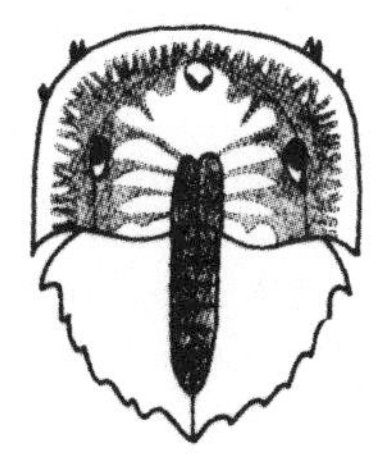

LARVAL STARFISH

pluteus stage

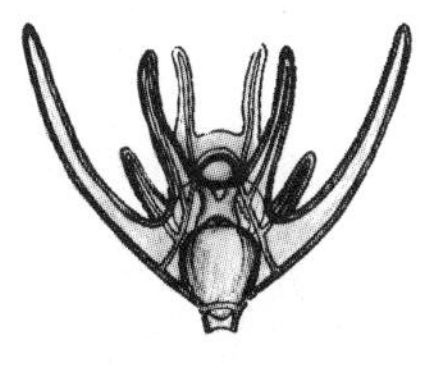

LARVAL CRAB

megalops stage

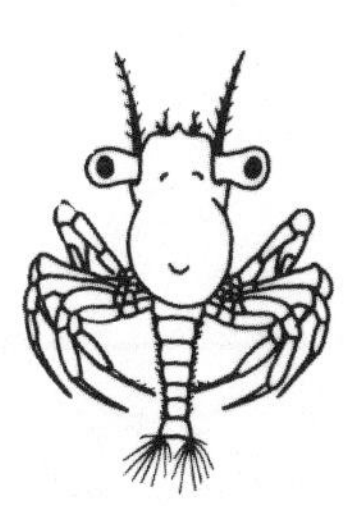

zoea stage

LARVAL SEA CUCUMBER

auricularia stage

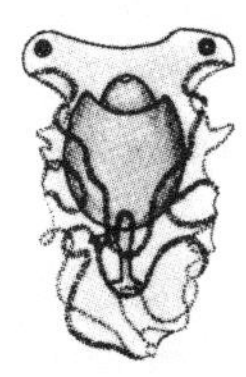

Larger, free floating algae

Although large, rooted plants cannot survive in the crashing surf, unattached seaweeds may be swept onto the beach to become part of the community strand line. Large masses often pile up after ocean storms.

SARGASSUM SEAWEED. *Sargassum* spp.—is a plentiful alga, having free-floating species and others that attach to solid underwater objects. Berrylike air bladders keep sargassum afloat.

Larger animals

The following animals, all coelenterates, also are planktonic in that they cannot move against a current but must float along with the water. They are much larger than zooplankton, and some in North Carolina waters grow to 20-25 cm across. Among the most beautiful of marine animals, most should be admired only from a distance because they can inflict a painful sting. At certain seasons, usually summer, these jellyfish "swarm" in coastal waters, with as many as five per cubic meter of water.

MOON JELLY. *Aurelia* sp. — is translucent, milky white or yellowish brown. Numerous short tentacles fall like a fringe around the body's margin. Moon jelly drifts through the water, trapping plankton in a mucous coating on the undersurface.

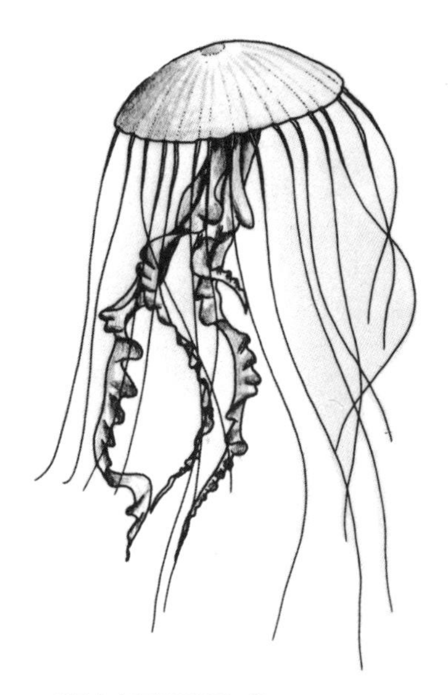

SEA NETTLE. *Crysaora quinquecirrha*—has about 40 long tentacles which possess stinging cells; drifts through water, paralyzing and consuming small animals.

LION'S MANE JELLYFISH. *Cyanea* sp.—is a common "red" jellyfish having a purplish red color. It may attain 1.5 meters in size in colder areas but is generally 0.3 to 0.7 meter across in North Carolina waters. Long tentacles hanging from the body may extend 15 meters and may number nearly a hundred.

PORTUGUESE-MAN-OF-WAR. *Physalia physalia*—has tentacles, 9 meters or more long, equipped with stinging nematocysts to paralyze and ensnare small prey, and is dangerous to humans. The animal is actually a colony of individuals acting together as a single organism. The "balloon" part of the body which floats on the surface may be purple, pink, blue or green, with a reddish ruffle across the top. DO NOT TOUCH!

The following animals are bottom dwellers. Unable to swim, they use hundreds of minute tubelike "feet" on the under surface to move over the sandy substrate. The kinship of sand dollars to starfish is evident in the five-rayed pattern on the body of each. In the sand dollar the rays appear to have become joined.

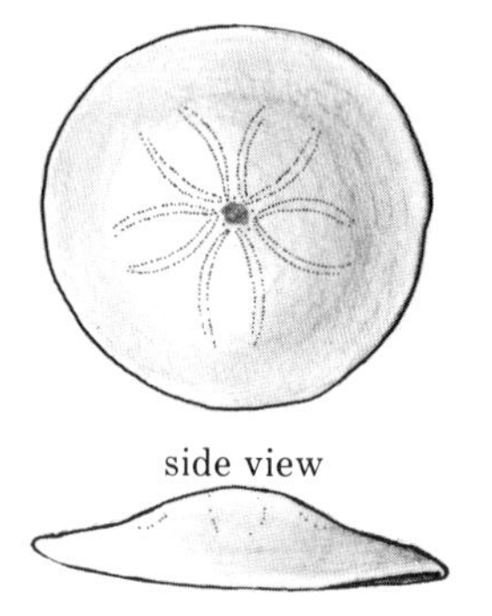

ATLANTIC SAND DOLLAR. *Echinarachnius parma*—like other sand dollars, eats minute organic particles located in sandy substrates. A series of small spines and cilia carries food particles to the mouth on the underside of the body.

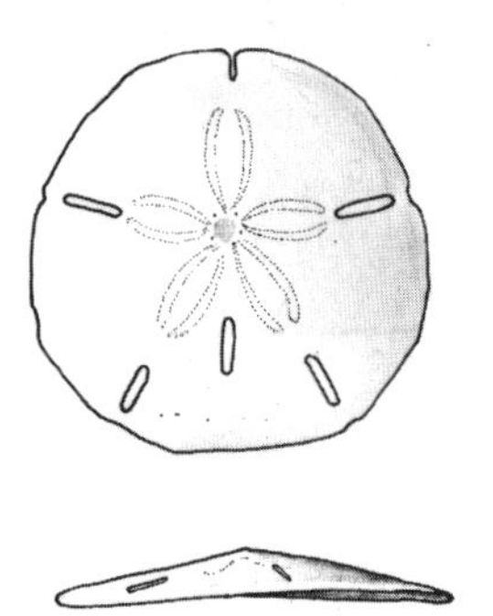

KEYHOLE URCHIN. *Mellita quinquiesperforata* — derives its common name from the "keyhole" shaped openings in its body.

STARFISH. *Astropecten articulatus*—has shorter purple arms, bordered by orange marginal plates. Starfish prey on small bivalve molluscs and other animals.

STARFISH. *Luida clathrata*—moves rapidly on long, slender arms, gray on top and cream colored beneath. Generally found beyond the heavy surf zone, starfish may be swept up near the beach.

The following heavyshelled molluscs burrow rapidly for protection against the rough sand and surf. Gastropods, having one shell, and pelecypods, with two shells, are abundant here.

Cockles, clams, arks, and discs are pelecypods. Their shells open for feeding on plankton, and a muscular "foot" extends for rapid digging.

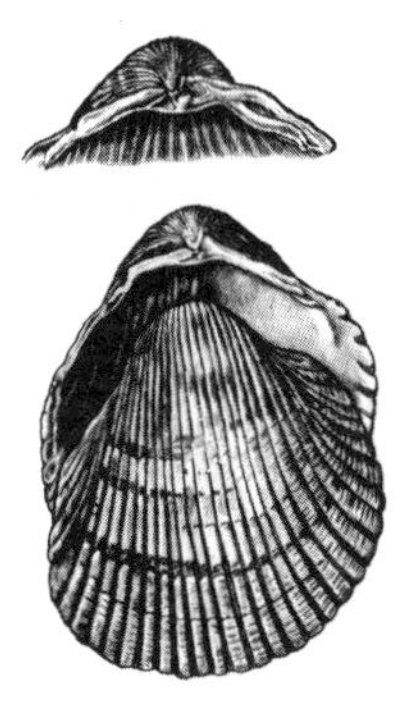

GREAT HEART COCKLE. *Dinocardium robustum*—is one of the biggest pelecypods in the beach community. Water passes over gills inside the body and flows through a mucous device where planktonic particles are caught and passed to the mouth.

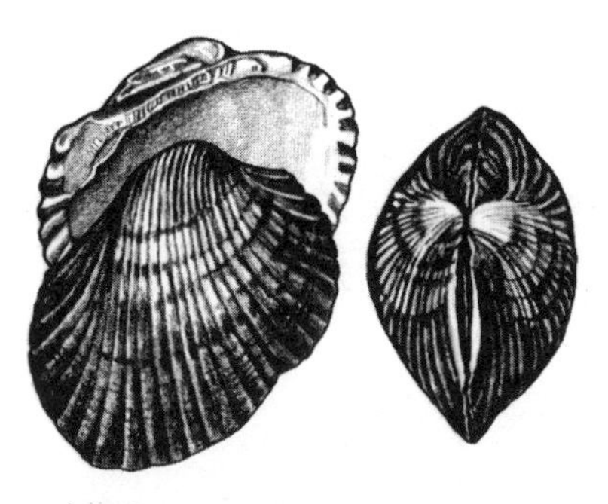

ARK SHELL. *Anadara ovalis*—is a suspension feeder with a mucous straining device for filtering plankton.

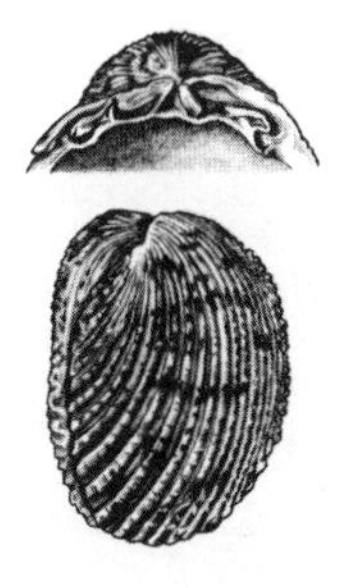

PRICKLY COCKLE. *Trachycardium egmontianum*—is smaller than the heart cockle and more elongated, with rows of small prickly spines along the shell. The prickly cockle also has an internal mucous net to catch plankton.

SURF CLAM. *Spisula solidissima* —may be very large (25 cm) when mature, and is one of the longest bivalves on the Atlantic coast. Short siphons draw in plankton which are then caught on the mucous net inside.

DISK SHELL. *Dosinia discus*— digs several centimeters into the sand before extending long siphons above the substrate surface to bring currents of water containing plankton to the mucous device inside.

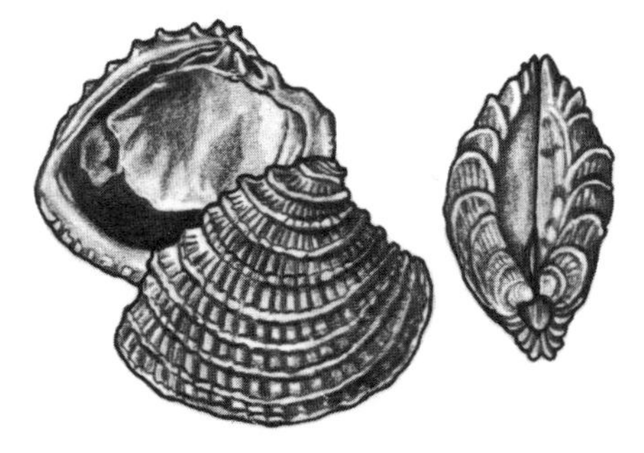

CROSS-BARRED VENUS CLAM or DOG CLAM. *Chione cancellata*— is marked with slightly raised ribs running both down and around the shell. This small species (5 cm) commonly occurs on the sand surface among eelgrass blades. It has short siphons for filtering planktonic food which is then caught on a mucous net.

Whelks and olive shells are gastropods which prey on pelecypods. A whelk grips its prey with its strong, muscular foot and wedges its shell between the two shells of the prey, prying the valves apart. Pulling with the foot, the whelk then forces the bivalve open. An olive shell has in its mouth a device called a radula which is similar to a shoeshine strap with many teeth on it. Pulled back and forth, the radula drills a hole in the prey's shell, and as the prey's muscles weaken the shell gapes open, exposing its inner contents.

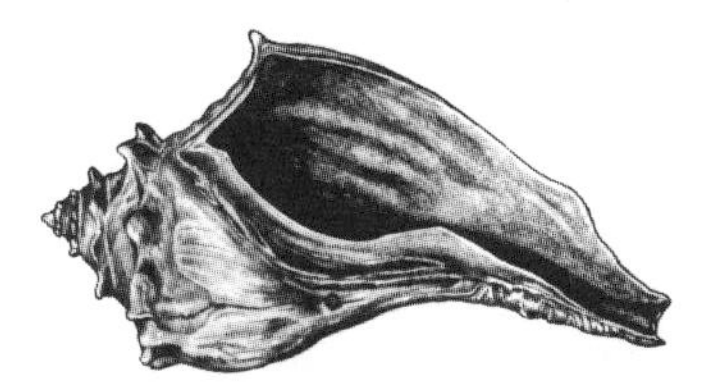

KNOBBED WHELK. *Busycon carica*—can be identified by holding the shell so that the top of the spiral points to the sky. In this species the shell opens on the right and the knobs are on top.

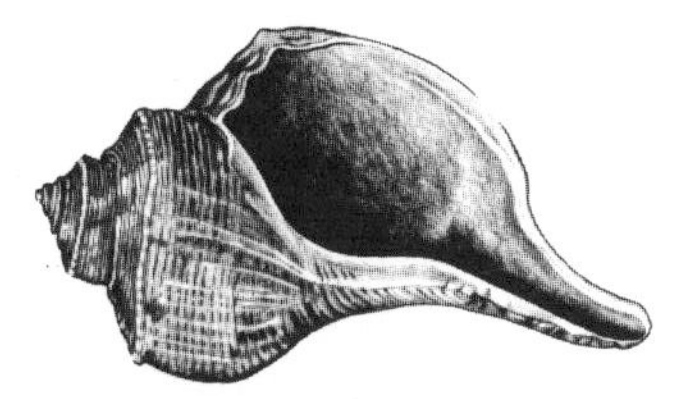

CHANNELED WHELK. *Busycon canaliculata*—lacks knobs on the shell top and the opening is on the right.

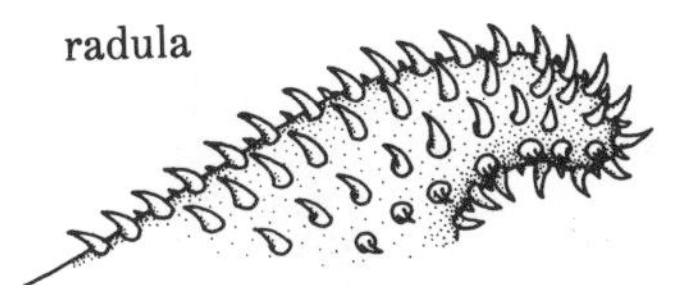

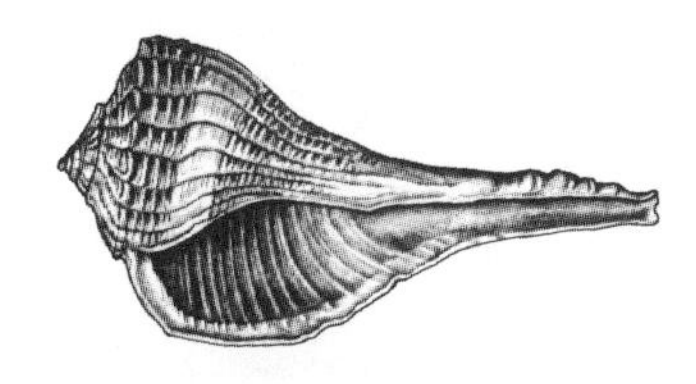

LIGHTNING WHELK. *Busycon contrarium*—has knobs on the spire, but the shell opening is on the left.

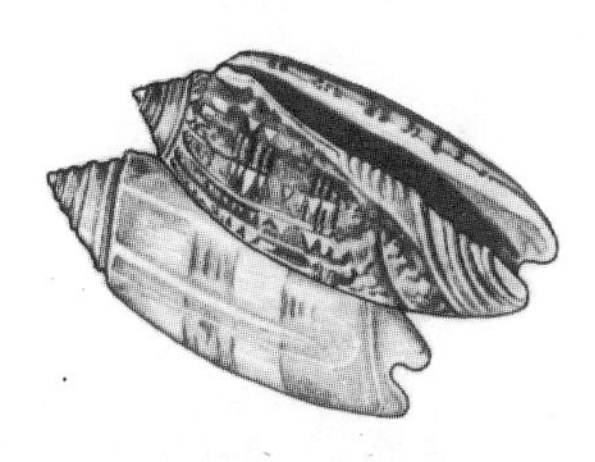

OLIVE SHELL. *Oliva* spp.—often has elaborate designs similar to "script"; thus some are called "lettered olive shells".

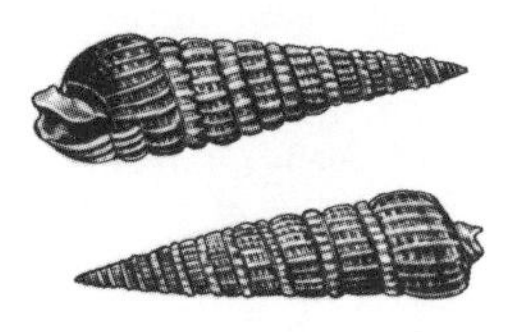

AUGER SHELL. *Terebra* spp.—is a gastropod which has a poison gland associated with the proboscis for paralyzing prey, usually polychaete worms.

Many crabs and shrimps are excellent swimmers. Swimming crabs have back legs modified into "paddling" structures, and shrimps have many legs that work effectively as "swimmerettes." Both crabs and shrimps burrow quickly into the sandy bottom.

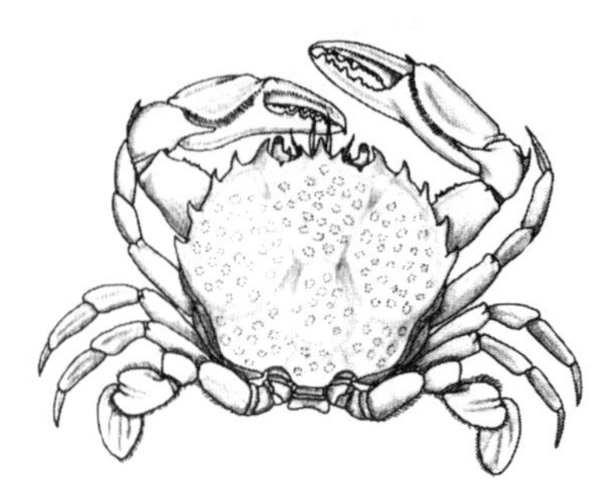

LADY CRAB or CALICO CRAB. *Ovalipes ocellatus*—has a nearly oval body with ornate decorations resembling a calico pattern. It scavenges on almost any dead animal encountered but is a good swimmer and can catch living prey as well.

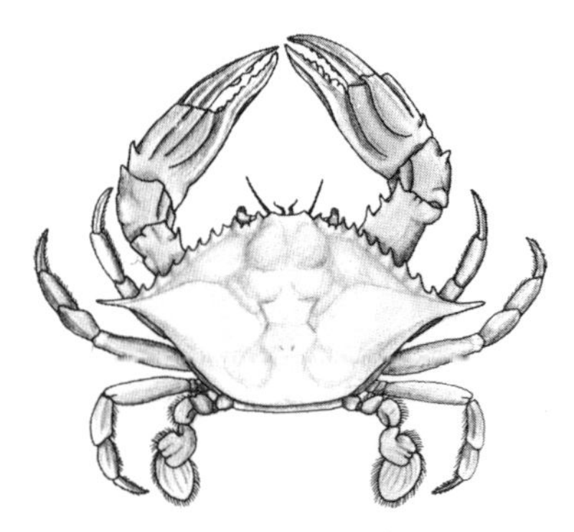

BLUE CRAB. *Callinectes sapidus* —occurs both in the estuary and the ocean during its life cycle. After the adults spawn in the ocean, larvae are swept through inlets into estuaries, the nursery grounds for developing young. This species feeds on dead animals and also pursues and consumes live ones.

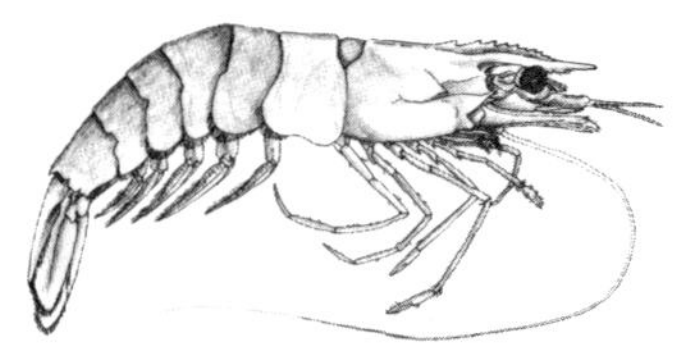

SHRIMP. *Penaeus* spp.—the shrimps most commonly eaten by people in North Carolina. Three different species occur in North Carolina: brown shrimp, *P. aztecus*; spotted or pink shrimp, *P. duorarum*. and white shrimp, *P. setiferus* These species spawn in the ocean, but the immature stages develop in estuaries where they feed primarily on detritus. As adults shrimps are scavengers and food finders, feeding on a variety of plants and animals, including their own kind.

GOOSENECK BARNACLE. *Lepas anatifera*—floats into beach waters aboard large pieces of driftwood to which it attaches with a flexible, extendable "neck." It is a filter feeder, using featherlike cirri to sweep the water and trap planktonic particles.

Most of the following vertebrates are active predators, placed higher in the marine food web, although some are plankton feeders.

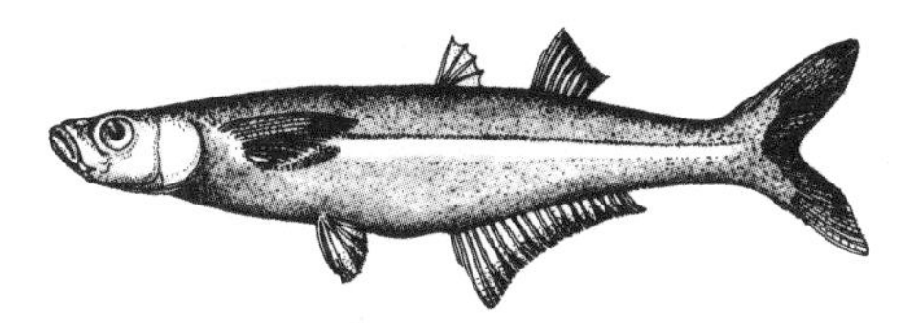

SILVERSIDES. *Menidia menidia*—is small, generally 7 to 13 cm long, and characterized by a horizontal silver streak along the sides of the body. The silversides feeds by darting through the water and filtering plankton through its mouth and gills.

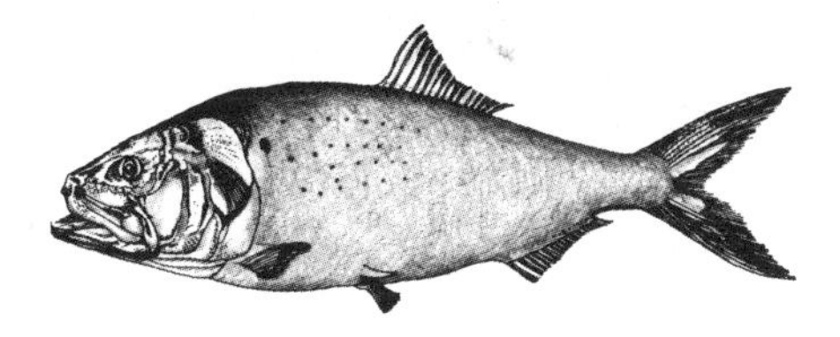

MENHADEN. *Brevoortia tyrannus*—swims with its mouth open so that water passes over the gills which strain up to 24 liters of water a minute. Plankton are caught in "rakers" on the gills.

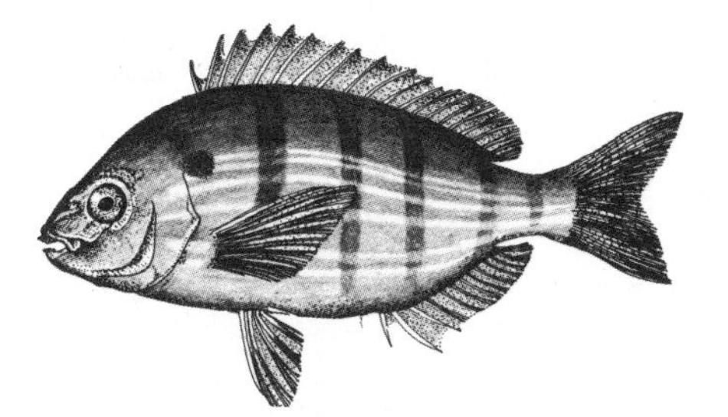

PINFISH. *Lagodon rhomboides*—is common in coastal waters, feeding on crustaceans, fishes, worms, shellfish, and seaweeds.

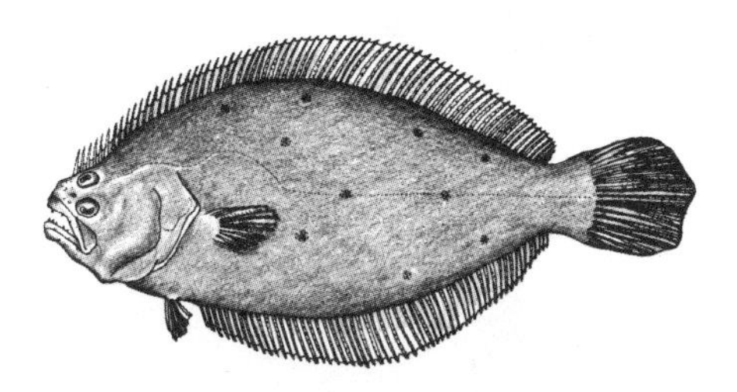

FLOUNDER. *Paralichthys* spp.—preys on small fishes and invertebrates, including shrimps. Several species are found here, each being most abundant at certain seasons of the year.

BLUEFISH. *Pomatomus saltatrix*—is a voracious predator on other fishes, particularly menhaden.

MACKEREL. *Scomberomorus* spp.—moves through schools of menhaden and other smaller fishes in a frenzy of feeding activity that froths the water and leaves large numbers of dead prey. The uneaten remains drift to the bottom where they are consumed by numerous scavengers.

LOGGERHEAD TURTLE. *Caretta caretta*—lives in the ocean, but the female moves to the warm sand beach to deposit eggs in summer. It feeds on shellfish, molluscs, jellyfish, and dead fishes.

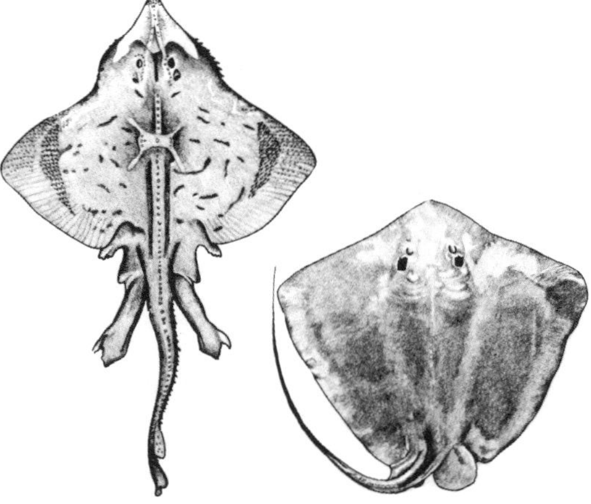

STING RAY. *Dasyatis* spp., and SKATE. *Raja* spp.—glide through the water and over the bottom, feeding on annelids, crabs, shrimps, squid, shellfish, and small fishes such as silversides. Both types have strong teeth for grinding their food.

Intertidal life

Plants

There are no permanent standing plants in the intertidal zone. Abundant phytoplankton, brought into this zone with the advancing tides, serve as the primary plant material for intertidal animals.

Animals

Few but the hardiest of species are able to live here, but these species, such as the following, occur in great numbers.

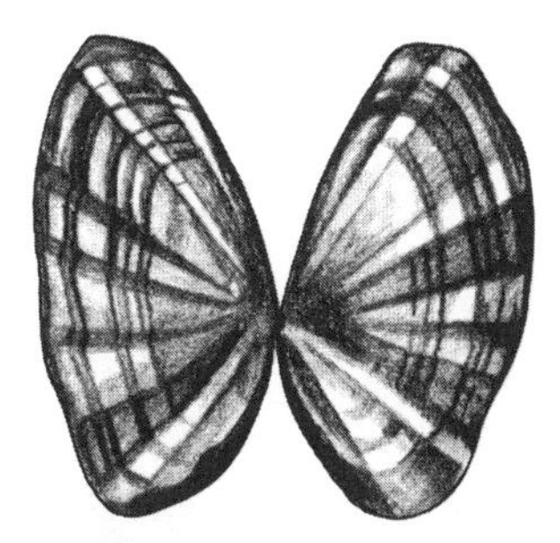

COQUINA CLAM. *Donax variabilis*—usually less than 2.5 cm long, occurrs in dense patches of hundreds per square meter. Each clam has a color pattern slightly different from all others. Burrowing rapidly into the sand as a wave recedes, the clam is swept out as a new wave advances. Like most clams the coquina strains plankton with a mucous device.

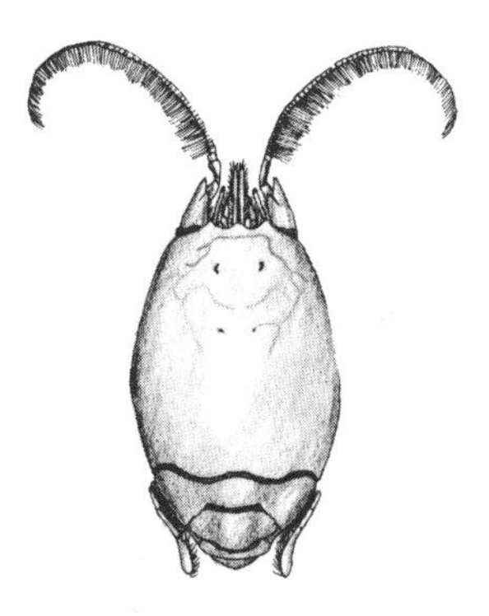

MOLE CRAB. *Emerita talpoida*—is a fascinating animal which burrows backward in the shifting sand each time a wave advances. This crustacean, 3 to 4 cm long, orients itself to the outgoing wave so that its antennae can filter the water for suspended plankton.

Many minute forms, called interstitial life or psammon, exist between the moist sand grains on the beach. Interstitial life includes various types of worms and arthropods.

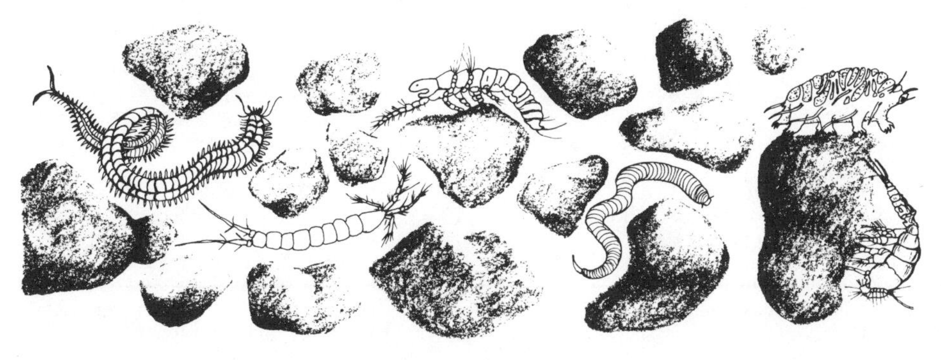

Birds fly in and out of coastal marine communities, and on the ocean beach fill a variety of niches.

COMMON TERN. *Sterna hirundo*—like other terns, hovers and plunges into the water to catch fish. Terns on the North Carolina coast range in size from the tiny Least tern, *S. albifrons*, to the full-sized Royal tern *Thallasseus maximus*, illustrated in the dune community. Terns feed almost exclusively on small fishes but occasionally eat sand eels and surface crustaceans.

SANDERLING. *Calidris alba*—like the numerous other species of sandpipers inhabiting coastal communities, this one runs to the water as an incoming wave flushes tiny crustaceans out of the sand. A sharp beak and keen eyes enable the sandpiper to pick these animals from the beach.

HERRING GULL. *Larus argentatus*—is one of the most important scavengers of the winter sand beach, along with the Ring-billed gull, *Larus delawarensis*. The blackheaded Laughing gull, *Larus atricilla*, replaces the Herring gull in summer. Gulls scavenge on almost anything and are important in cleaning beaches and waters of decaying refuse. They occasionally dive into a wave from 5 meters above the water to catch a live fish.

Supratidal life

High tides deposit plant and animal debris at the extreme upper tide line. This debris provides food and protection for organisms in the community. Many insects and amphipods thrive in this stranded material.

Plants

This arid stretch of sand, lying only a few meters from the ocean, is usually uninhabited by plants. Adjacent to the sand dune community, the ocean beach may support an occasional plant from the frontal dune area.

Animals

Few species, most of them arthropods, can withstand the intense heat and light on the upper beach.

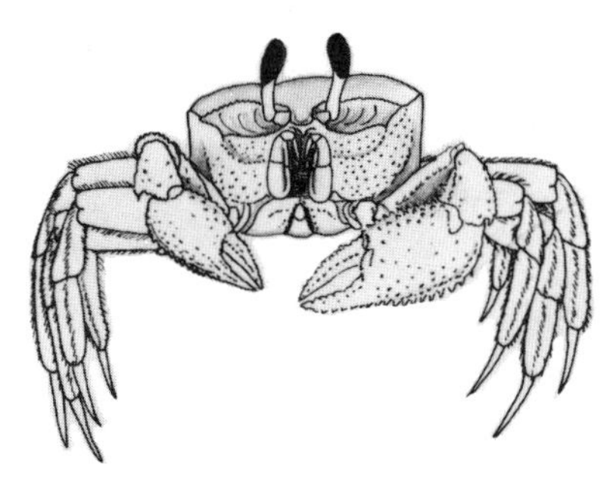

GHOST CRAB. *Ocypode albicans*—is highly specialized for living in this zone. It excavates deep tunnels with front and back entrances and seldom leaves the burrow by day. At night it scavenges on dead and decaying matter washed onto the strand line. Although this crab appears to live in a "desert," it still is essentially a marine organism. It breathes through gills which must be wet with ocean water daily. The female deposits its eggs in the ocean, and the young develop there.

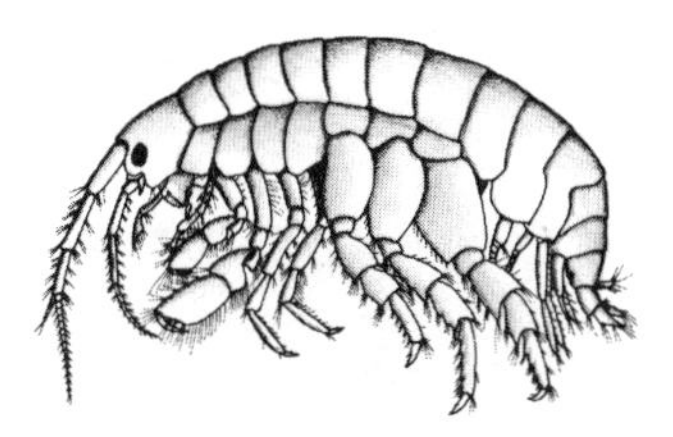

SAND HOPPER or BEACH FLEA. *Talorchestria* spp.—is an amphipod which avoids extremes of the habitat by seeking shelter in small burrows or under seaweed debris. It scavenges on dead plant and animal matter on the beach.

SAND DUNES

Sand Dune Habitat

The habitat

Sand dune habitats occur wherever loose, shifting sands build into mounds called dunes. Coastal dunes usually are found adjacent to the ocean beach, which provides most of the dune sand. Dunes also are found on"spoils islands" located in sounds behind the barrier islands where people have pumped in mounds of dredged sand and silt. In time the barren mounds become vegetated with hardy dune plants which help stabilize the sand.

The row of dunes nearest the ocean is directly exposed to the full effects of ocean beach environmental forces — strong winds, whipping sand, and glaring sun. Ocean water floods the dunes only during extreme storms. Behind frontal dunes the habitat is more protected and different types of plants, including maritime forest shrubs and trees, grow on the second and third rows of dunes.

Special features

—Harsh physical forces characterize the dune habitat. Strong winds are a nearly constant force the year around, with prevailing winds from the southwest in summer and northeast in winter.
—Winds sweep quantities of sand particles back and forth daily, causing dunes to shift and migrate, often covering vegetation.
—Winds carry salt spray from the ocean to the dunes. The salt spray, heaviest during storms, periodically kills back the tips of shrub and tree branches facing the ocean. Behind frontal dunes salt spray causes such plants to grow stunted and to appear sculptured.
—The water supply on frontal dunes is scarce. Water quickly seeps down from the surface, seeking the natural water table which may be 6 meters or more below.
—Salt spray provides the main source of nutrients on frontal dunes. Nutrient minerals leach out of the sand quickly, producing low fertility. Nitrogen is particularly scarce in dune soil since little or no decaying plant and animal matter accumulates to enrich the habitat.
—Intense light is reflected by the bright sand surface which is extremely hot in summer and cool in winter. Below the surface, temperatures decrease considerably.
—On back row dunes, canopies of shrubs and trees provide some cover which moderates the effects of wind and sun. Surface temperatures and evaporation are reduced, increasing soil moisture.
—Oxygen is plentiful in this habitat.

Adaptations of plants and animals

—Frontal dune plants have flexible blades able to withstand whipping about by wind.
—The roots of many dune plants grow downward and outward rapidly, keeping pace with sand accumulation. A year's growth, represented by the distance between nodes on the roots, may be 30 cm or more.
—On the dunes, where the sun causes intense heat and light, plants have adaptations which prevent rapid loss of water from plant tissues. The leaves curl inward and turn to a vertical position to decrease exposed surface. Many leaves are thick with a waxy coating that reduces evaporation. Pores of leaves may close at midday to prevent loss of fluids.
—Dune plants have extensive and/or deep root systems to avoid surface heat and evaporation and to reach the water table deep below the dunes.
—Dune animals often burrow or dig tunnels into the sand to avoid extreme surface temperatures and to gain protection from predators.
—Many insects and mammals forage at night to avoid intense daytime heat and light.
—Dune animals often have very light coloration which reduces the amount of heat absorbed by the body. The ghost crab appears even lighter in the day than at night, probably partly for better camouflage.
—Some animals produce additional body water by oxidation during metabolic processes. This is important where water supply is inadequate.

The community

Dunes commonly are said to support a "desert of the beach" community. The habitat appears to be sparsely supplied with the basic necessities for life. Nevertheless, the few species adapted to living in this habitat, particularly in the frontal dune area, are highly specialized and quite successful. They often occur in large numbers. Plants, which dominate the dune community, are the first organisms to invade the habitat. As plants vegetate the sand, animals follow and the dune community forms.

Inorganic nutrients and water are at a premium, especially in the frontal dune area, and plant productivity is relatively low. Cycling of organic to inorganic material, so prevalent in decomposition processes in other nearby communities, is greatly reduced on frontal dunes. Thus, dune plants largely depend on salt spray, borne in the air from the crashing surf, as the primary source of inorganic nutrients. Salt spray falls on the blades and leaves of dune plants or settles on dune surfaces. Inorganic materials and water contained in the spray quickly leach downward and percolate into the sandy soil. Dune plants send roots deep into the sand to catch moisture and absorb nutrients. The roots form a matted growth that often riddles the interior of a dune, helping to hold and stabilize the sand. Many dune plants also absorb moisture and nutrient materials directly from salt spray collected on the surface of their blades.

The food supply for dune animals is relatively sparse compared to that of other communities. Some scavenging animals eat dead and decaying plant and animal debris that accumulates at the strand line. The ghost crab and certain insects forage on this stranded material, particularly at night. Other animals feed on available dune plants and in turn are eaten by predators. Since little or no organic detritus accumulates on the frontal dune substrate, animals that need this type of food cannot settle there.

Behind the frontal dunes and on the back rows of dunes, less resistant plants are able to survive. Salt spray, which may be harmful to leaves, is less intense. On shrubs and trees, foliage facing away from the ocean receives less spray than foliage facing the ocean and grows quite well.

Where rows of dunes are protected from salt spray and beach forces by frontal dunes, the community changes. There, dune thickets and maritime forest trees, much like those of inland wooded areas, can survive. Under the canopy of shrubs and other trees water is retained in the soil. Organic detritus accumulates under the trees and decomposes to inorganic nutrients used by shrubs and trees. Woodland animals such as rabbits, mice, and birds become part of the community.

Typical organisms

Life on front dunes

Plants

Larger rooted plants are the producers on the ocean side of the dune. Particularly tolerant of strong winds, sun, and salt spray, these plants are called "pioneers" since they are the first to vegetate a barren dune.

SEA ROCKET. *Cakile* spp.—grows in clumps with numerous stalks. Like the sea elder it has fleshy leaves, but their edges are more sharply toothed. The flowering bodies resemble tiny rockets.

SEA OAT. *Uniola paniculata*—covers frontal dunes, especially their tops, and has extensive root systems that reach 1.5 to 10.0 meters below the surface for water. This is the primary dune-building and stabilizing plant on the mid-Atlantic coast. The long, narrow leaves curl to prevent loss of water in the windy, arid habitat.

AMERICAN BEACH GRASS. *Ammophila breviligulata*—is not native south of Cape Hatteras, but has grown well when planted in southern areas. The grass holds sand effectively and aids in dune building.

SEA ELDER. *Iva imbricata*—is a 0.6 to 1.0 meter tall shrub with bright green succulent leaves. Elder grows in scattered clumps over the whole dune. A cousin to the marsh elder, it is able to withstand heavy salt spray and open beach forces.

view looking down on plant

DUNE SPURGE. *Euphorbia polygonifolia*—covers the ground like a mat in places. Branches radiate from a single root and have short, narrow leaves that resemble a doily. When punctured the branches exude a milky fluid.

PANIC GRASS. *Panicum amarum*—stands out among other frontal dune plants as a stately clump, 1.0 to 1.5 meters tall and with outsized leaves.

In many instances the following plants exist on the frontal dune but are more abundant on the back side of front dunes and on the back rows of dunes.

SEASIDE EVENING PRIMROSE. *Oenothera humifusa*—grows as a low spreading mat with slightly fuzzy, pale green leaves. The delicate, yellow cuplike flower grows on small stalks.

CROTON. *Croton punctatus*—has whitish-gray leaves, oval in shape. The leaves have a strong, fragrant odor when crushed.

Animals

Few animals are adapted to living on the frontal dune. Resident animals, mostly insects, often forage on the upper beach area adjacent to the dunes.

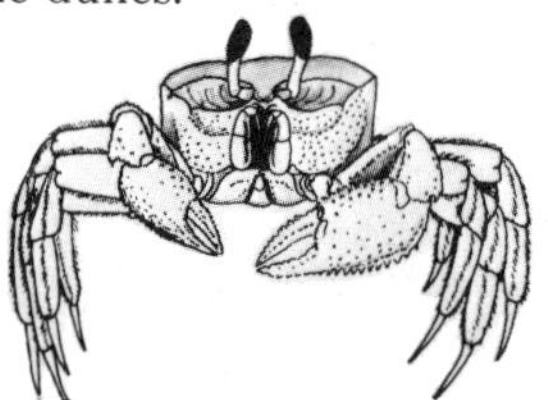

GHOST CRAB. *Ocypode albicans*—scampers across the beach, moving sideways on spindly legs. It has a very light coloration that makes it relatively inconspicuous and prevents much absorption of heat. This crab digs elaborate burrows to avoid day-time heat and scavenges on beach refuse primarily at night. Eyes waving on long stalks provide 360 degree vision.

MOLE CRICKET. *Gryllotalpa hexadactyla*—is a vegetarian, feeding on young roots and seedlings of dune plants under which it burrows in the daytime.

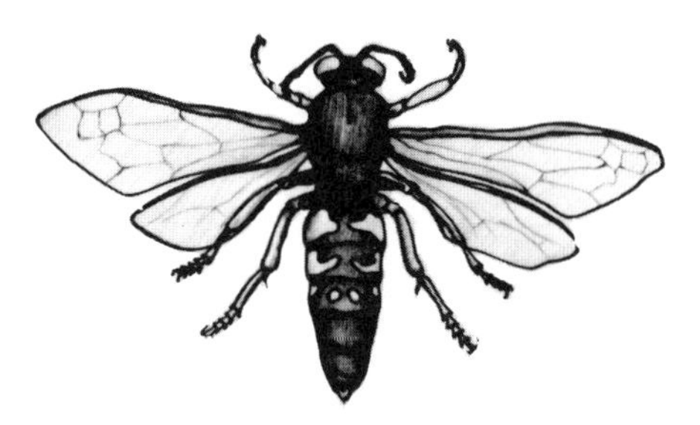

DIGGER WASP. *Bembex* sp.—burrows to cooler sand. It frantically digs, then flies up to cool its body. The wasp covers the burrow to protect its young from predators and parasites. It preys on flying insects, pulling them into its nest to nourish the larvae.

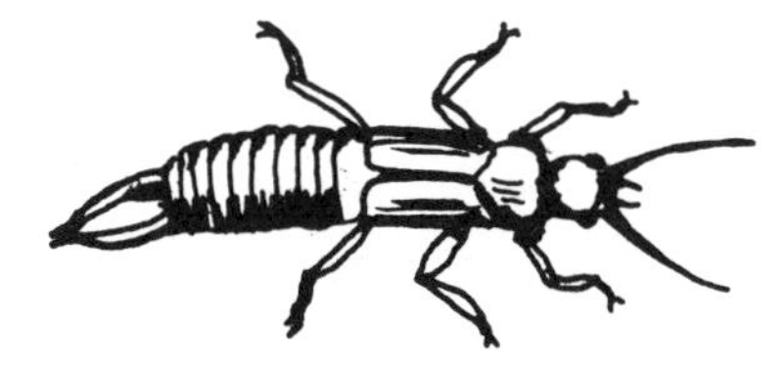

EARWIG. *Anisolabis* sp.—is nocturnal, seeking shelter under beach debris by day. It scavenges on beach and dune refuse and may attack the larvae of other insects or feed on succulent foliage.

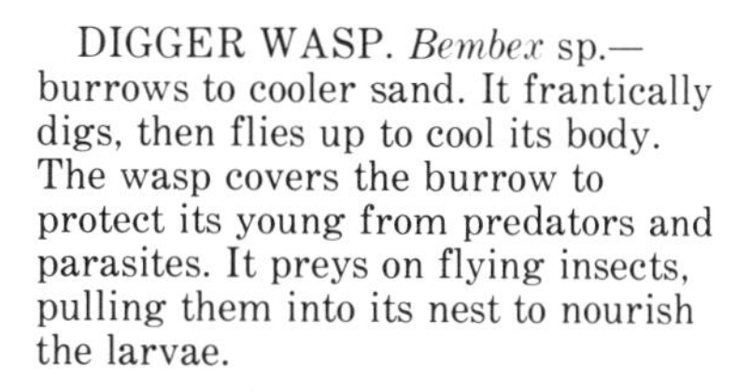

SAND HOPPER or BEACH FLEA. *Talorchestia* spp.—is an amphipod which leaps when its strand line hiding place is disturbed. It scavenges the remains of seaweeds and dead animals washed onto the beach, and its bite causes an itching sensation in human intruders.

VELVET ANT. *Dasymutilla* sp.—is actually a wingless wasp, fiery red, which can inflict a painful sting. The body is densely furred for insulation. The female lays eggs in the nest of the digger wasp where the hatchlings feed on the wasp's larvae.

Life on back dunes

Plants

The type of vegetation behind frontal dunes depends on the elevation of the sand. (see Figure 9).

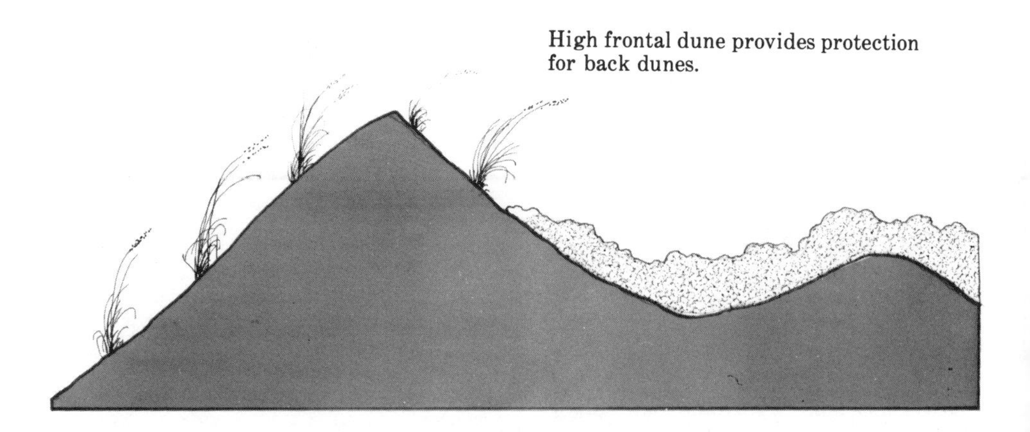

Figure 9. Relationship between dune height and type of dune vegetation.

If dunes on the second row are higher than those in front, the ocean side of the back dunes suffers the full force of ocean winds, with heavy salt spray and sand blasting. Such dunes support only plants characteristic of frontal dunes. In areas where higher frontal dunes reduce the effects of wind-whipped sand and salt spray, additional types of plants and meadowlike grasses, such as the following, occur.

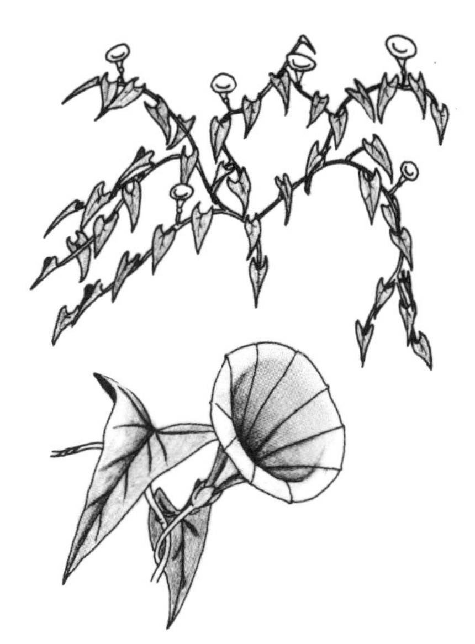

BEACH MORNING GLORY VINE. *Ipomoea* spp.—looks like the typical morning glory, running as a vine along the ground. Its small pinkish flowers resemble trumpets.

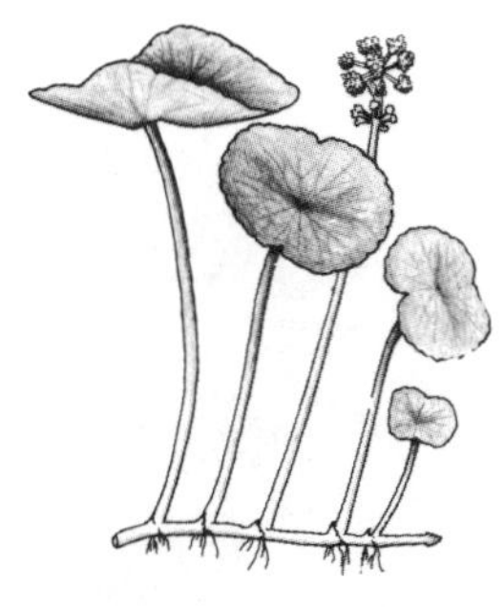

PENNYWORT. *Hydrocotyle bonariensis*—blankets the sand with bright green, waferlike leaves the size of a half-dollar. Small erect stalks, topped by a single leaf, rise at regular intervals from dense underground runners.

BROOMSTRAW RUSH. *Andropogon* sp.—flourishes behind protective frontal dunes. Easily recognized, it stands out with heads like broomstraw and gives the landscape a rusty hue.

SANDSPUR. *Cenchrus tribuloides*—is the well-known prickly "sticker" that is hard to miss on a barefoot hike through the dunes.

CATBRIAR. *Smilax* spp.—weaves around and among other dune shrubbery. Blanketing other plants with its intertwining vines, it reduces wind damage to dune shrubs.

SALT MEADOW HAY. *Spartina patens*—often the predominant meadowlike grass, grows sparsely on frontal dunes, but it usually grows more densely behind them. When covered by sand the grass quickly recovers, growing rapidly to its former height.

SPANISH BAYONET. *Yucca* spp.—derives its name from its stiff, swordlike leaves. It often grows 3.0 meters tall.

Shrubs and shrubby forms of trees also grow on back dunes protected by higher frontal dunes (see Figure 9). Dense thickets, pruned by salt spray into prostrate, spreading forms, include such species as yaupon, *Ilex vomitoria*; red cedar, *Juniperus virginiana*; live oak, *Quercus virginiana*; sweet bay, *Persea borbonia*; and beach olive, *Osmanthus americanus*.

YAUPON. *Ilex vomitoria*—often exhibits the effects of salt spray when new growth, budding on the side facing the ocean, is "burned" by accumulating salt. New buds die, leaving only barren twigs. Thus, the yaupon appears sculptured, as if the wind had permanently bent the branches back.

Protected from ocean forces by high frontal dunes, a "maritime" forest oftens forms behind low-growing shrub and tree thickets. Trees in the forest grow taller than those in thickets just behind the front dunes. Beneath the forest canopy develops an understory of American holly, *Ilex opaca*, and Hercules' club or toothache tree, *Zanthoxylum clavaherculis*. Sometimes dogwood, *Cornus florida*, and loblolly pine, *Pinus taeda*, grow upward to form part of the canopy. Catbriar and wild grape weave thick vines from the forest floor to reach sunlight at the canopy top.

Animals

Some dune animals move back and forth between dunes and maritime forest. They get water from freshwater ponds in the wooded area but can move to the dunes for scavenging and foraging at night.

MEADOW MOUSE. *Microtus pennsylvanicus*—burrows beneath dune shrubs and thickets, venturing out at night to feed on dune grasses.

COTTONTAIL RABBIT. *Sylvilagus floridanus*—forages on dune grasses and tender young shoots of vegetation at night.

Some birds forage in the dunes and a few species nest there in grasses as well as in shrubs and trees. The Painted bunting, Catbird, Mockingbird, Brown thrasher, and Cardinal are the most common nesters in the shrub thickets.

In winter large numbers of blackbirds sometimes linger on the dunes, and sparrows are common at that season.

CATBIRD. *Dumetella carolinensis*—eats insects such as beetles and grasshoppers, as well as wild grape and holly berries in season.

PAINTED BUNTING. *Passerina ciris*—has a variety of food sources including insects, wild fruit, and the seeds of weeds. The Painted bunting leaves the thickets and woods of barrier islands and flies south for the winter.

MOCKINGBIRD. *Mimus polyglottos*—prefers a diet of insects and berries similar to that of the Catbird, and frequents the dune thickets to feed.

CARDINAL. *Richmondena cardinalis*, state bird of North Carolina—consumes a variety of insects, wild grapes, holly berries, and blackberries.

BROWN THRASHER. *Toxostoma rufum*—grasshoppers and beetles are the thrasher's main food sources, along with wild fruits and acorns in the dune thickets and maritime forest.

Dunes can form wherever obstructions catch loose sand and may occur in situations other than on the ocean front. Sometimes the action of wind and water causes dunes 7 to 10 meters high to form on islands in the sounds and on dredge piles that dot the margins of artificially deepened channels. The sand usually is coarse and mixed with broken shells. These dunes are taken over, in time, by pioneer plants, and in a hundred years may bear tall trees.

Before the bare dunes become vegetated to any extent, they often are occupied by nesting colonies of water birds. Natural camouflage is provided there for the mottled, sand-colored eggs laid in sand-hollow nests. When disturbed, nesting birds rise up, screaming their protests. Often they divebomb the intruder to protect their nests. Water birds feed in a variety of communities. The most common species include the following.

LEAST TERN. *Sterna albifrons*—flies over the water, apparently feeding exclusively on marine life and hovering briefly before diving to capture small fishes and crustaceans. The Least tern is only 23 cm long.

COMMON TERN. *Sterna hirundo*—feeds on small fishes such as pipefish and immature menhaden, on planktonic crustaceans, and occasionally on insects.

CASPIAN TERN. *Hydroprogne caspia*—when feeding, flies with its beak pointing downward, close to the water's surface where it plunges to capture fishes. This bird also fishes while sitting on the water. In addition, the Caspian tern will eat the eggs and young of other birds.

GULL-BILLED TERN. *Gelochelidon nilotica*—like other terns the Gull-billed tern dives into the water for small marine animals, but more often it walks about and stoops to pick up prey. Fishes, crabs, and worms are favorite foods, and this tern also likes frogs, lizards, small mammals, and particularly insects such as the marsh grasshopper.

BLACK SKIMMER. *Rynchops nigra*—feeds on surface dwelling fishes and crustaceans. Flying low, it skims the water with the lower half of its beak, then circles back to pick up organisms driven to the surface by the disturbance.

OYSTER CATCHER. *Haematopus palliatus*—crabs, worms, snails, limpets, cockles, mussles, clams, and oysters all are part of this bird's diet. The Oyster catcher inserts its bill into the partially opened shells of molluscs and cuts the prey's muscles so that the shell gapes open.

WILSON'S PLOVER. *Charadrius wilsonia*—is also called the Thick-billed plover because of its long, thick black bill. Small crustaceans such as crabs and shrimps, as well as small molluscs, and insects are its food.

SEMIPALMATED PLOVER. *Charadrius semipalmatus*—uses shallow-water flats and marshes as its feeding grounds during low tide, and it also feeds among sandpipers at the water's edge. The Semipalmated plover scampers about while feeding, stopping suddenly to grab marine worms, small crustaceans, and molluscs from the sandy bottom.

ROYAL TERN. *Thallasseus maximus*—feeds primarily on small fishes, diving into the water from considerable heights to capture its prey. It fishes both offshore and in the sounds. The Royal and Caspian terns grow to 59 cm and 50 cm, respectively, dwarfing the tiny Least tern.

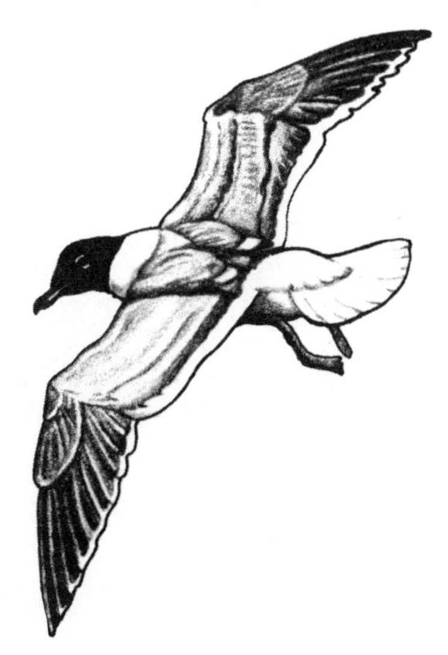

LAUGHING GULL. *Larus atricilla*—is a scavenger on shore-line debris, but also catches fishes and shrimps in the shallows of tidal flats as well as dragonflies and other insects in flight.

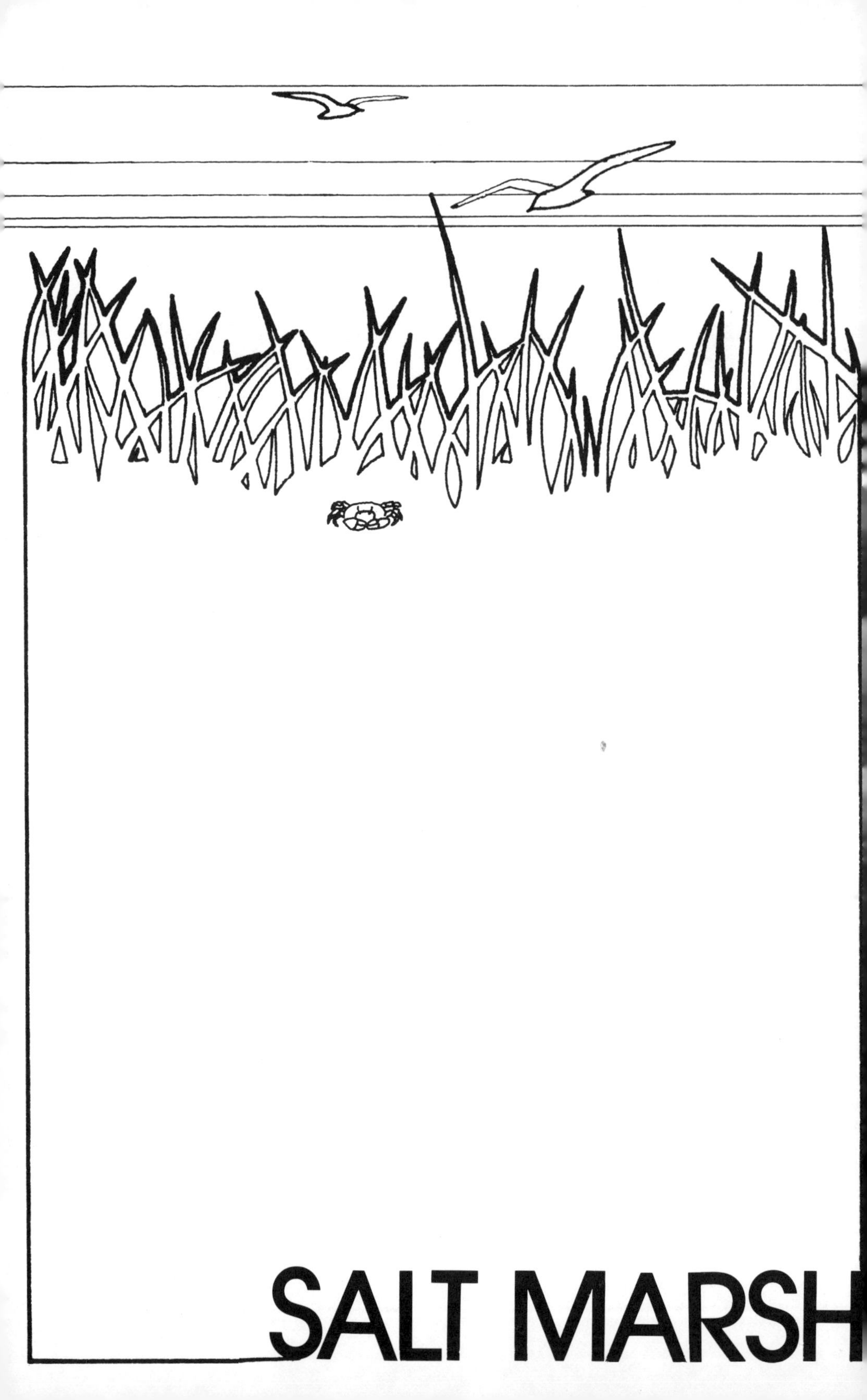

SALT MARSH

Salt Marsh Habitat

The habitat

Salt marshes edge the sound-side shorelines of North Carolina's barrier islands and mainland. They extend hundreds of miles along the intertidal shore of sounds, estuaries, creeks, and rivers where the water is essentially salty. Although plants are abundant in salt marsh habitats, only a few species make up the greater part of marsh vegetation. The primary plant is salt marsh cordgrass, which is adapted to a salty, aquatic existence. The substrate is formed of mud and silt precipitated from low velocity waters and caught among the plants. With time, sediments may accumulate, elevating the habitat and making it no longer intertidal. Other plants then invade the habitat and a new type of community eventually develops.

Special features

—Marshes form in extensive intertidal areas. The habitat gradually slopes down to the subtidal zone. There may be a drop of 0.3 meter or more at the low tide line at what is called the marsh bank.

—Waves and winds striking the salt marsh are usually moderate to gentle, but at times may be quite strong. The thick covering of grasses reduces their effects.

—Ebbing and flooding tides may cause wide changes in salinity.

—The marsh is covered by shallow water at high tide and exposed to air at low tide. Evaporation may leave heavy salt residues in the substrate and cause excessive salinities in the water.

—Tidal currents usually leave marsh water turbid and full of silt.

—Surface temperatures of water and substrate may change rapidly as cool, incoming tidal water floods sunheated marshes.

—The substrate is muddy. Except for occasional shell remains or wood debris, solid substrate is virtually absent in the marsh. Intertwining roots and stems of marsh grasses form a dense mat at and beneath the surface.

—The oxygen supply is poor in water over the marsh and beneath the mud surface. Bacterial decomposition of organic material depletes the oxygen and leaves the mud rich in hydrogen sulfide, which gives marshes a characteristic "rotten egg" odor.

Adaptations of plants and animals

—Many marsh organisms can tolerate significant salinity changes; some can withstand extremes from very salty to almost fresh.
—Cordgrass tolerates high salt concentrations in water and soil. On hot days the plant exudes salt crystals that heavily coat its blades. Marsh glasswort withstands a very high salt content without stress.
—Certain snails climb stalks of the taller intertidal marsh grass to prevent being covered by water at high tide. Other snails can withstand the high salt content and temperatures of small tidal pools left on the marsh at low tide.
—Competition is fierce among attaching animals for any solid substrate. Mussels attach to the base and roots of marsh grasses. Barnacles and microscopic algae attach to the grass blades. Oysters cluster on top of one another.
—The dominant marsh grass is anchored to the substrate by strong, extensive roots that bind mud and silt together.
—Many organisms are adapted to making good use of limited oxygen in the substrate. Some crabs increase oxygen below the surface by digging burrows.
—Many organisms feed on plentiful detritus deposited on the surface or suspended in the water.
—Some species consume foods not actually in the water. Snails, for instance, graze on minute algae growing on blades of marsh grass.

The community

Because it has a wide variety of plant and animal species occupying a large number of niches, the salt marsh community is one of nature's most self-sustaining ecological systems and is a valuable coastal resource. Decomposition of marsh grass and debris produces rich supplies of inorganic nutrients essential to abundant plant growth and renewal of the marsh grass itself. The marsh community supplies virtually all its own needs rather than being dependent on other communities.

Although few animals feed on living cordgrass, tides flush large quantities of grass debris into the waters where it is broken down to detritus. It can then support large numbers of zooplankton or provide food for filter-feeding animals such as oysters, clams, scallops, fishes, and shrimps. The larval stages of these species and many others use detritus as their major source of energy. Deposit feeders and sediment ingesters obtain this food from the substrate. Suspension feeders extract detritus as well as plankton from the water. Organisms that cannot digest such material feed on bacteria and fungi embedded in the detritus.

Typical organisms

Subtidal life

Plants

Phytoplankton are plentiful because of the abundance of inorganic nutrients. At high tide phytoplankton are washed onto the muddy surface of the marsh where they are trapped and become food for deposit feeders and sediment ingesters. (See the Ocean Beach section for examples of phytoplankton.)

Few large algae or grasses survive in the subtidal marsh, where settling mud and silt would smother them. Occasional algae seen there probably have broken off and washed in from nearby communities.

Animals

Dense populations of zooplankton float through the waters over and around the marsh and serve as food for plankton feeders. The falling tide leaves plankton stranded and they become part of the rich sediments deposited on the substrate. (See Ocean Beach section for examples of zooplankton.)

Estuarine waters flowing through the marshes are nursery grounds for larval and juvenile stages of many marine animals. Here immature stages can survive in calm waters, nourished by plentiful food and protected by marsh grasses. Included are such animals as the croaker, spot, menhaden, flounder, blue crab, squid, and shrimp. Most adults are able to move freely back and forth between different marine communities. (See other communities for illustrations.)

The following animals are subtidal dwellers, moving onto the marsh at high tide. They commonly dwell in tidal creeks which riddle the marsh.

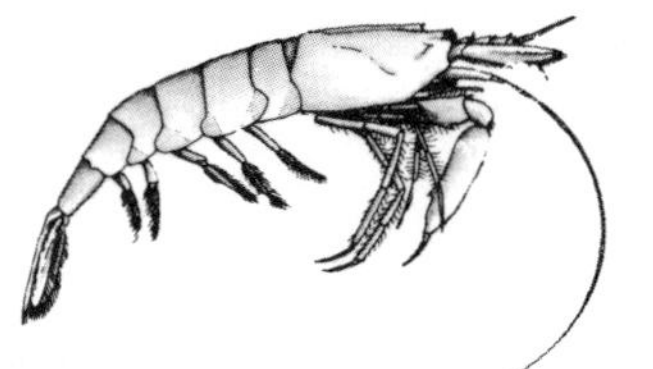

SAND SHRIMP. *Crangon semtemspinosa*—moves from tidal streams into the marsh at high tide, consuming small bottom-dwelling invertebrates as well as plankton.

BLUE CRAB. *Callinectes sapidus*—scavenges almost any dead material and is an active predator, catching small, swimming crustaceans and fishes.

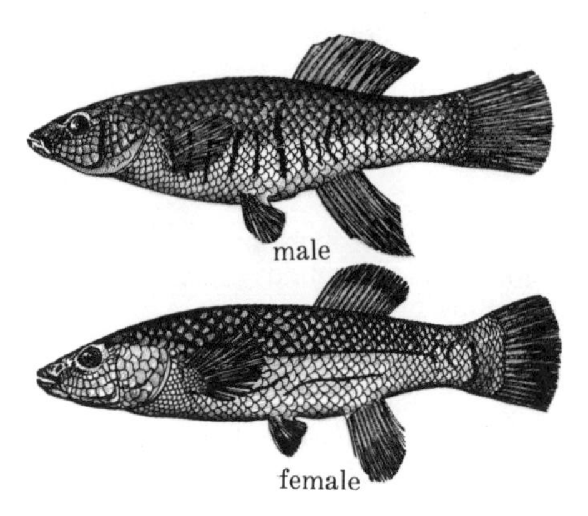

KILLIFISH. *Fundulus* spp.—may grow to 13 cm long. It is abundant in most coastal marine communities and feeds on a variety of small invertebrates. Male and female body patterns differ slightly.

Intertidal life

A blanket of bacteria, fungi, and algae covers the entire intertidal marsh substrate. Algae are producers; bacteria and fungi are decomposers. In concentrated patches they often color the substrate pink, brown, yellow, or purple. Tiny algae also live on blades of marsh grasses. Beneath the substrate multitudes of anaerobic bacteria, which do not require oxygen for metabolism, continue the process of decomposition. All serve as important and plentiful sources of organic nutrients for other animals.

Larger plants

The characteristic appearance of the marsh is due to the vast amount of large, visible plants occupying the intertidal zone.

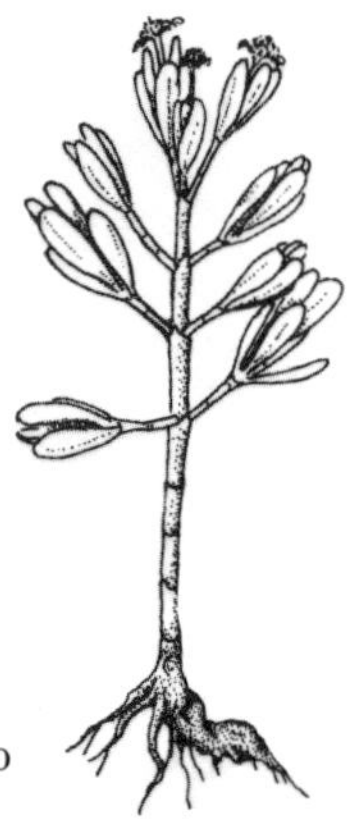

SEA OXEYE. *Borrichia frutescens*—is generally found in the upper intertidal zone, and may grow wherever elevated spots occur. It has thick, whitish-green leaves and bright yellow composite flowers similar to small sunflowers.

BLACK NEEDLE RUSH. *Juncus roemerianus*—has tall (1 meter) needlelike blades and is dark gray. It grows in the upper intertidal zone or in other elevated areas covered by salt water only during unusually high tides. Needle rush completely replaces cordgrass as the ecologically dominant plant in more elevated marshes along the North Carolina coast.

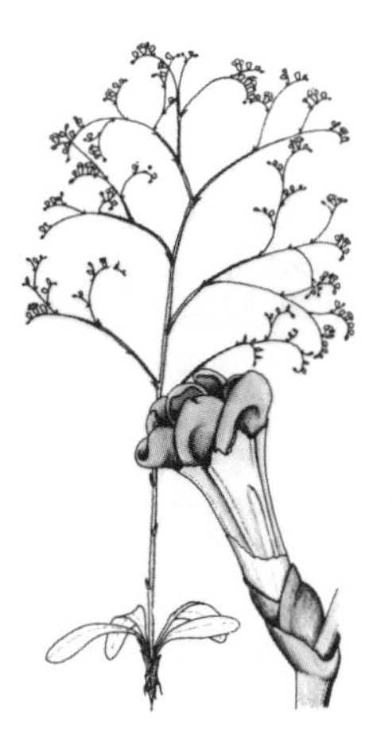

SEA LAVENDER or MARSH ROSEMARY. *Limonium carolinianum*—grows at the fringe of the upper intertidal marsh. This delicate plant has large lancelike leaves at the base, with small stems extending upward. Tiny lavender flowers bloom in summer and fall.

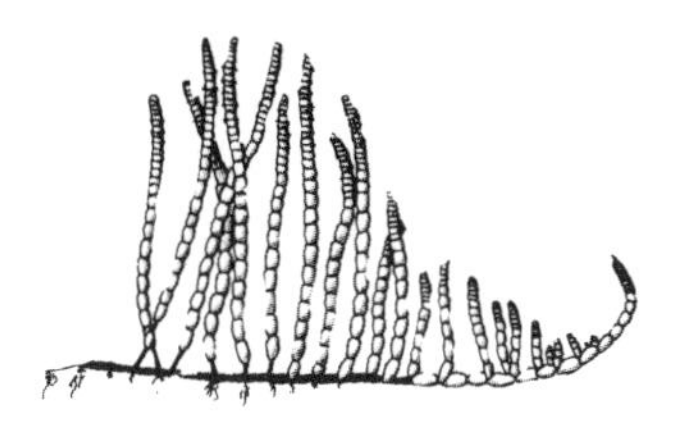

MARSH GLASSWORT. *Salicornia virginica*—may extend throughout the intertidal zone, intermingling with cordgrass, and may grow on salt flats in the marsh. It is salt tolerant and may have a salt content greater than that of the surrounding water. Glasswort looks like long green pipe-cleaners with jointed stems. There are three species in coastal marshes. The marsh samphire, *S. europa*, turns pink in fall.

SALT MARSH CORDGRASS. *Spartina alterniflora*—is the most abundant and ecologically most important large plant of the marsh. It grows tall near the water's edge (from 0.6 to 1.2 meters high), but is stunted in elevated areas. Although it dies back in winter, its roots remain alive and give rise to new growth in spring. Cordgrass provides the bulk of detritus to the community.

Animals

Some species are permanent residents, while others move to and from the marsh with the tide. Many marine worms burrow or move freely through the marsh mud, contributing to its aeration. The following animals are permanent residents. Generally found at lower elevations near the water, they dwell among the blades and roots of cordgrass, in the mud, and along the marsh bank.

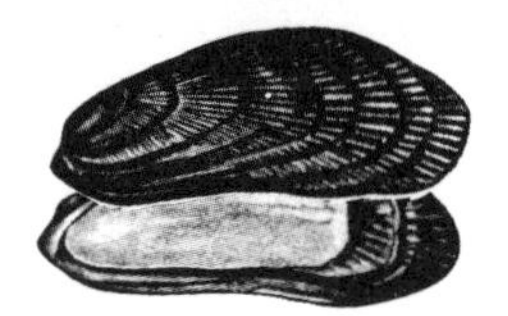

RIBBED MUSSEL. *Modiolus demissus*—is a suspension feeder with a mucuous device for straining plankton, bacteria, and organic detritus suspended in the water. Like the oyster, the ribbed mussel attaches to any solid object available in the lower intertidal zone.

EASTERN OYSTER. *Crassostrea virginica*—is the common edible oyster species in North Carolina. It attaches to the base of cordgrass or scattered shell remains in the lower intertidal zone. The shells open at high tide when the oyster feeds on plankton filtered from the water, but close tightly at low tide or when the animal is exposed to very low or high salinities. When the animal is subjected to low salinities for several days its body volume may increase.

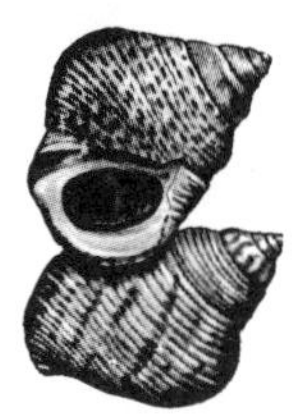

MARSH PERIWINKLE. *Littorina irrorata*—is a small snail which climbs stalks of cordgrass when the tide rises, since it would drown if submerged for long. The periwinkle uses its radula, a filelike structure of the mouth, to rasp algae from cordgrass, rocks, shells, and driftwood and to remove the film of algal vegetation from the muddy marsh surface.

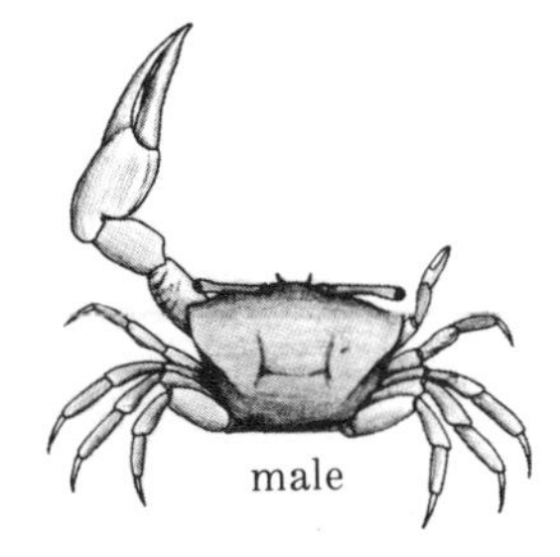

MARSH FIDDLER CRAB. *Uca pugnax*—is a well-known shoreline resident which looks very much like its close relative, the sand fiddler crab, *U. pugilator*. The marsh fiddler prefers a muddier substrate than the sand fiddler. Both dig burrows into the substrate; the marsh fiddler, among the densely matted roots of cordgrass; the sand fiddler, on salt flats or sandier upper intertidal areas. Both species feed by scraping organic matter from the substrate. The male has one large and one small claw and the female has two small claws.

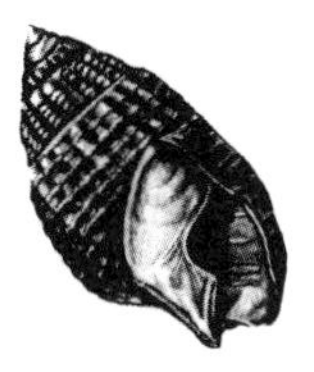

MUD SNAIL or SMOOTH BASKET SHELL. *Ilyanassa obsoleta* — is the most abundant intertidal gastropod on the Atlantic seaboard. It is found in dense clusters of hundreds in the muddy marsh, moving slowly over the substrate, consuming detrital food, microscopic algae, and the eggs and larvae from egg cases of worms. The snail also scavenges the remains of fishes, shrimps, and molluscs.

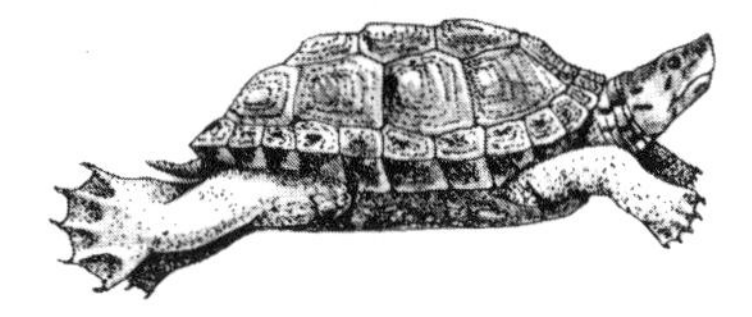

DIAMONDBACK TERRAPIN. *Malaclemys terrapin*—feeds on dead fishes, live fiddler crabs, and small molluscs in the marsh.

Supratidal life

Plants

The following plants, found in the upper reaches of the marsh, generally are flooded with salt water only during storm tides. Some may extend slightly into the upper intertidal area. Such plants, located above the influence of tidal waters, contribute little or nothing to marsh food webs. Birds often nest among their blades.

EASTERN BACCHARIS. *Baccharis halimifolia*—has thin, light green leaves, toothed near the tip, which are shed in winter. The plant is most noticeable in autumn when flowers resembling tiny cotton balls cover the bush, providing the common name of "cotton bush." It is highly resistant to salt spray and can withstand occasional flooding by salt water.

MARSH ELDER. *Iva frutescens*—competes with baccharis and myrtle for space at the raised margins of the marsh.

SPIKE GRASS. *Distichlis spicata*—is a short, wiry grass growing with and among salt meadow hay.

WAX MYRTLE. *Myrica cerifera*—is a dark green shrub with tough, leathery leaves. It often grows in dense clumps and has a distinct bayberry aroma.

SALT MEADOW HAY. *Spartina patens*—is a close relative of cordgrass, but cannot exist in the intertidal zone. Salt meadow hay is used as forage for livestock in many coastal areas and forms widespread, meadowlike expanses of grass above the high tide line.

SEASIDE GOLDENROD. *Solidago sempervirens*—has stems 0.6 to 1.2 meters tall, and long, tapering leaves. In late summer and fall the characteristic bright yellow composite flowers of goldenrod form dense clusters on the stem.

Animals

Some animals living in the supratidal area move down to the marsh to feed, becoming an integral part of the community.

MARSH CRAB. *Sesarma reticulata*—resembles a fiddler crab but has a noticeably rectangular carapace, long spindly legs, and two tiny claws. It can be found easily by lifting its protective covering of strand line material. The crab burrows at the upper reaches of the marsh. It eats dead animals, attacks small living animals such as wounded fiddler crabs, picks algal film off rocks and driftwood with its claws, and feeds on the outer leaves of cordgrass.

COFFEE BEAN SNAIL. *Melampus lineatus*—inhabits the upper marsh, generally at and above the tide line. This tiny snail resembles its namesake and occurs in dense numbers, particularly beneath boards and decaying plant material in the strand line. It is considered a "transition" species, changing its life habits from aquatic to terrestrial, but its eggs still must be deposited in water. The snail feeds on detritus and small algae.

RACCOON. *Procyon lotor*—feeds in the marsh at low tide and has sharp claws which enable it to dig mussels and crabs.

DRAGON FLY. *Erythrodiplax berenice*—a carnivorous insect, it eats other insects and is sometimes called "mosquito hawk".

MEADOW MOUSE. *Microtus pennsylvanicus*—feeds on marsh vegetation and often nests among the salt meadow hay.

Many birds are associated with the marsh. Some visit just to feed; others nest among the grasses.

CLAPPER RAIL. *Rallus longirostris*—is more commonly heard than seen as it slips through the intertidal marsh grasses seeking small fishes, marsh crabs, fiddler crabs, aquatic insects and snails. It builds nests of cordgrass stems and leaves at the upper fringe of the tidal zone.

MARSH WREN. *Telmatodytes palustris*—quick moving and secretive, this wren is seldom seen. It catches marsh insects, and sometimes it nests on the fringes of the marsh.

REDWING BLACKBIRD. *Agelaius phoeniceus*—eats marsh insects. Nests of this species, made of marsh grasses, are found in upland borders.

SEASIDE SPARROW. *Ammospiza maritima*—nests among salt meadow hay along the borders of most sounds and estuaries of North Carolina. It moves into the marsh to feed on insects and onto nearby tidal flats for additional feeding.

WILLET. *Catoptrophorus semipalmatus*—nests on the ground among salt meadow hay. It is very vocal, producing a frenzy of screaming when any intruder approaches. The Willet feeds on molluscs and crustaceans during low tide on the marsh and on bordering flats.

BOAT-TAILED GRACKLE. *Cassidix mexicanus*—is a large, irridescent blackbird that eats both live and dead plant or animal material in the intertidal zone.

MEADOWLARK. *Sturnella magna* —visits the upland fringes of the marsh for feeding. Its diet includes marsh insects.

The following herons are summer residents which nest in large colonies in North Carolina marshes. Some, like the Black-crowned Night Heron and the Great Blue Heron, also winter there in smaller numbers. Herons feed in the shallow water in and around marshes, consuming fishes, insects, crustaceans, and other small animals which they capture with their spearlike beaks.

BLACK-CROWNED NIGHT HERON. *Nycticorax nycticorax*—grows to 0.7 meter tall but appears shorter because of the habit of folding the neck so that it rests on the body. Like other herons it feeds on fishes, crustaceans, insects, and frogs.

GREAT BLUE HERON. *Ardea herodias*—is the largest of the herons shown here, standing 1.2 meters tall. The body is bluish gray, but the crown, bordered by black feathers, is white. The Great blue heron feeds on small fishes, eels, snakes, mice, and even on small birds in the marshes and on tidal flats.

SNOWY EGRET. *Leucophoyx thula*—is the smallest of the herons shown here, growing to 0.6 meter tall. Its body is white with yellow feet and black bill and legs. The Snowy egret runs a zigzag pattern through the shallows in search of shrimps and crabs and feeds on insects and frogs also.

AMERICAN EGRET. *Casmerodius albus*—has a yellow bill and black legs and feet. The American egret grows to 1 meter tall and is smaller than the Blue heron. The egret stalks its prey slowly before striking its beak out for fishes, frogs, snakes, and a variety of other small animals.

LOUISIANA HERON. *Hydranassa tricolor*—wades in the water, pausing before spearing its prey which includes small fishes such as minnows and killifish, and shrimps. In the marsh, insects such as grasshoppers are its favorites. The white rump and under parts distinguish the Louisiana heron which grows to 0.7 meter tall.

TIDAL FLAT

Tidal Flat Habitat

The habitat

Broad mud or sand flats develop in quieter waters of North Carolina sounds, where suspended particles of fine grained sand and silty mud settle to the bottom and form vast expanses of sandy-muddy substrate. Tidal flats occur either as extensions of the shoreline or as shoals. They are basically intertidal, although the surrounding area covered continuously by water is part of this habitat also. Organisms adapted to living on sandy substrates may be quite different from those adapted to muddy substrates. However, many species occur on both substrates or move freely from one to another, so sand and mud substrates are considered here as a single habitat.

Special features

—Flats are broad expanses of land covered by shallow water at high tide and exposed to air at low tide.
—The intertidal habitat offers little protective cover and appears starkly barren compared to the nearby salt marsh.
—Fine sands are deposited where current velocity is relatively strong in sounds. Silt is deposited where currents are weak, usually farther up the estuary. Most tidal flat substrates, however, commonly consist of a mixture of sand and mud. Often muddy sediments are covered with a thin layer of sand. The substrate is more stable than that in some other habitats, with little shifting of sediments.
—The quiet waters are laden with fine mud and silt which can smother animals by clogging their feeding and breathing mechanisms.
—Tidal water prevents flats from drying out appreciably, even at the surface. Ripples in the sand and tidal pools retain water at low tide, increasing soil moisture.
—Solid substrate where organisms can attach is limited to occasional empty shells.
—Wave action, modified by the shallow water, is relatively gentle and wind effects are minimal.
—Dissolved oxygen is ample in the water but often is inadequate in muddier substrates. High summer temperatures in shallow water and underlying sediments lower the amount of dissolved oxygen.
—Water temperatures fluctuate considerably during the tidal cycle. In summer the sun heats flats during low tide, but the rising tide brings in cooler water. In winter the reverse occurs.
—Water over the flats can have a wide range of salinities. Evaporation causes high salinity in small tidal pools. The water may become less salty during heavy rains, or at ebb tide as brackish water flows out from rivers. Within hours, however, the flooding tide may return saltier water.

Adaptations of plants and animals

—Organisms readily survive alternate exposure to air and flooding by salt water twice daily.
—Few organisms remain on the exposed surface at low tide. Most burrow, retreat into tubes in the sand, or move with tides to the subtidal zone.
—Burrowing by organisms, particularly worms and molluscs, protects them against predators on the exposed flats. Many animals establish extensive permanent burrows or construct elaborate tubes in the substrate into which they can withdraw rapidly. Others build temporary sand tubes or move freely through the substrate.
—Many bivalve molluscs and worms have anticlogging mechanisms to clean the silt from their burrows and vital body surfaces.
—Most tube builders and burrowers have some means of moving water rich in oxygen, into their structures. The paddlelike swimmerettes of some burrowing crustaceans and the cilia and parapodia of certain worms are examples. Plankton and detritus suspended in the water also are drawn into the tubes.
—The tops of many types of worm tubes extending above the substrate surface are camouflaged with fragments of shells and bits of seaweed.
—Certain snails, such as the mud snail, burrow slightly into the substrate or congregate in small tidal pools to avoid desiccation at low tide. They are able to withstand wide fluctuations in both water temperature and salinity.
—Subtidal grasses, rooted in the substrate, are most common in sandier flats but also occur in muddy areas.
—Many small organisms dwell among the blades of subtidal grasses which provide them protective cover.
—Attaching forms cover the surface of shell remains. Oysters pile on top of one another or cluster on a single shell fragment. Certain seaweeds attach to the shells of living animals such as the dog clam.
—Beneath the oxygen-poor substrate thrive vast numbers of anaerobic bacteria.
—Plankton flow in with the rising tide, supporting a large number of filter-feeding animals.
—Many burrowing molluscs have long siphons which can be extended into the water above the substrate to draw in oxygen and food.
—Sediment ingesters and deposit feeders, adapted to feeding on the abundant deposited organic matter, are very common.
—Many of the numerous scavengers and food-finders on the flats locate their prey with well developed senses of "smell" (chemosensitivity).

The community

Tidal flat communities develop and flourish on the extensive mud-sand habitats. Tides and currents carry inorganic nutrients and detrital organic matter, often from nearby marshes, to the flats. Inorganic nutrients provide ample minerals for the community's plants and much of the organic detritus settles on the substrate, providing an important source of energy for animals.

On many flats, rooted marine grasses blanket the lower intertidal and subtidal sand, providing excellent protective cover and a home for myriads of immature and small animals. Seaweeds also attach to blades of such grasses. In addition to grasses and seaweeds, bountiful phytoplankton and zooplankton provide food for a variety of other feeders. As the tide rises additional types move over the flats and become part of the food web.

The community depends largely on the microscopic algae, bacteria, plankton and detritus which float suspended in the water or settle on the substrate. Many animals tunnel freely through the mud or creep over the surface to find and consume deposited material. Others that build tubes or burrows create a current of water to draw suspended food down into their hiding places. The food web of the tidal flat involves a greater variety of species than do food webs in most other communities. Thus, a larger number of feeding niches is occupied.

Typical organisms

Subtidal and intertidal life

Microscopic plants

Phytoplankton typical of other coastal waters occur in large numbers and are the main producers. In muddier areas, suspended silt often decreases penetration of light in the water, limiting the productivity of phytoplankton. (See Ocean Beach for examples of these.) Vast numbers of microscopic mud algae settle on the substrate.

Larger algae and grasses

Many large seaweeds occur, particularly in the subtidal zone, and some species extend into the lower and even the upper portion of the intertidal zone.

Green, brown, and red seaweeds are most abundant on the sandier portions of tidal flats, but they occur in fewer numbers than in the jetty community where there is more solid surface for attachment.

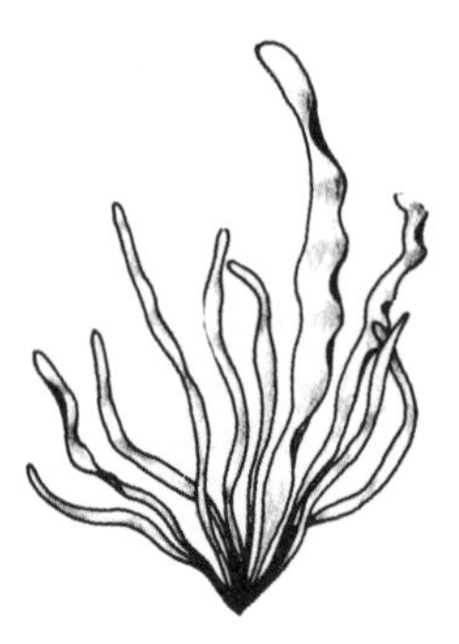

SEA HAIR or LINK CONFETTI. *Enteromorpha* spp.—similar to sea lettuce, is thin as tissue paper and bright green or yellow green. However, it grows in streamers like ribbons of confetti, some with flowing blades 30 cm or more long. The streamers of some species are very fine, resembling "mermaid's hair." Sea hair attaches by a holdfast to solid objects and grows in the subtidal zone and into the intertidal zone.

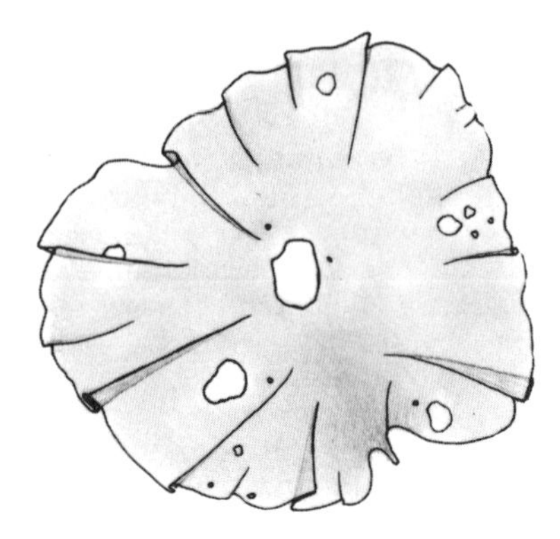

holes due to feeding of grazing animals

SEA LETTUCE. *Ulva lactuca*—as its common name implies, this alga has the appearance of silky, wilted, bright green lettuce. The frond is a broad, flat, thin sheet with slightly ruffled edges. Sea lettuce attaches to shells and rocks by a small disk, called a holdfast. The plant is common just at and below the low tide line but also extends into the lower intertidal zone.

The following species, found at and below the low tide line, occasionally can be seen in tidal pools in the intertidal zone where they remain covered by water at all times.

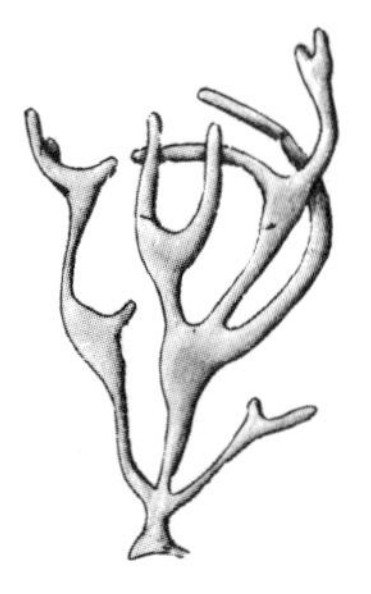

SPONGE SEAWEED or DEAD MAN'S FINGERS SEAWEED. *Codium* spp.—has long, rounded fronds up to 30 cm or more in length. Fingerlike shoots branch and divide from the central frond. When this dark green alga is pulled apart, the broken ends seal themselves immediately to prevent water loss.

COULTER'S SEAWEED. *Neoagardhiella* spp.—is similar to sewing thread seaweed. The irregularly spaced branches are slender, smooth, rounded and fleshy. Narrow at the base, they broaden slightly and taper at the ends.

HYPNEA. *Hypnea* spp.—is a red alga, but gray green pigments often dominate its color. Abundant in April, it usually is gone by June. The main stem has many irregularly spaced branches which produce additional branches. The ends of the long branches are "naked," bending like small hooks.

CHENILLE-WEED SEAWEED. *Dasya* spp.—is reddish purple and grows subtidally and most abundantly in spring. When floating in water its small rounded branches, densely covered with fine hairlike fringes, are delicate and feathery.

SEWING THREAD SEAWEED. *Gracilaria* spp.—varies in color from dingy purple to olive green. Its many thin branches, which feel like cooked spaghetti, do not tangle when lifted from water. The stem is short and rounded at the base but gets flatter and broader as it branches irregularly into "palmlike" segments which themselves similarly branch. Red algae such as this are used commercially to make agar.

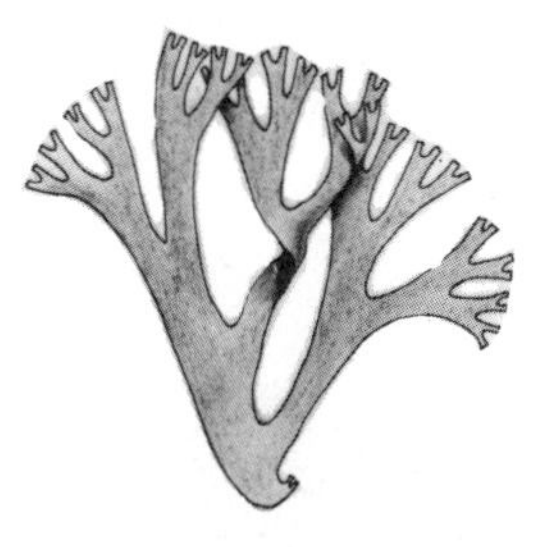

DICTYOTA. *Dictyota* spp.—looks like brown rockweed, but its olive brown fronds are flat like tissue paper and may fork repeatedly.

Subtidal grasses may extend into the lower intertidal zone. Few animals actually consume the living grasses. As the blades die during winter, however, they decompose into detritus and become an important source of energy for the community.

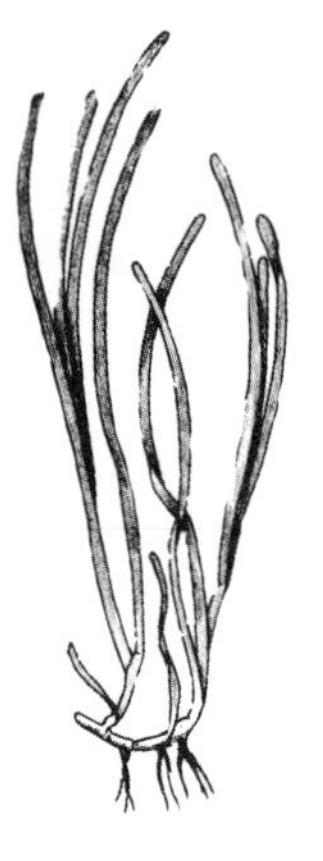

EELGRASS. *Zostera marina*—has roots, like terrestrial grasses, and is one of the seed bearing plants of the sea. It blankets the sandy subtidal bottom, extending at times into the intertidal zone as well. Eelgrass provides food for water fowl and protective cover for scallops, crabs, fishes, shrimps, and a wealth of tiny marine life among its blades. Seaweeds such as *Ectocarpus* attach to the grass. Widgeon grass, *Ruppia* sp., and shoalgrass, *Halodule* sp., are very similar to eelgrass, and occupy comparable subtidal areas.

Animals

The young stages of many marine animals develop among subtidal grasses which give them protective cover. The adult stages of molluscs, annelids, and crustaceans are particularly abundant on the flats. Many are burrowers, living near the sand surface in summer but burrowing more deeply in winter. They riddle the substrate with holes and mounds.

MOTTLED DOG WHELK. *Nassarius vibex*—occurs in sandier areas of the flats, thus avoiding direct competition with the mud snail. Both species have a proboscis for sucking in food and appear to feed on a variety of things. They cluster around egg cases of marine worms, eating the eggs and larvae. They also feed on microscopic algae and scavenge on dead fishes, shrimps, and bivalve molluscs. Both snails have a well developed sense of "smell" (chemosensitivity) for finding food.

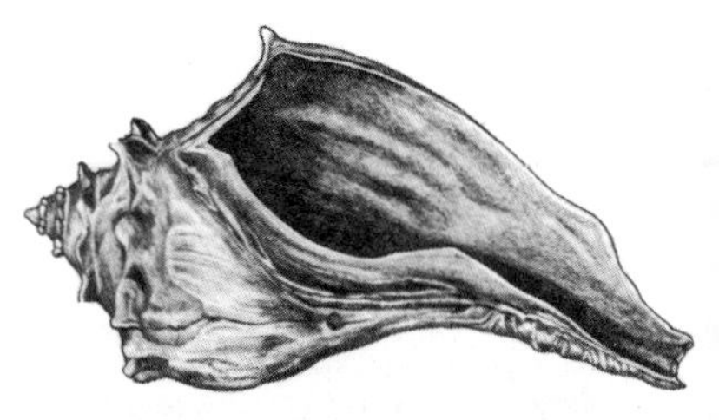

WHELK or CONCH. *Busycon* spp.—has a radula, but does not use it specifically for boring. Instead, the whelk wedges the thick outer lip of its shell between the two shells of a bivalve to pry open and expose the contents. The radula may be used to help scrape out the tissue. (See Ocean Beach for all three species of this genus.)

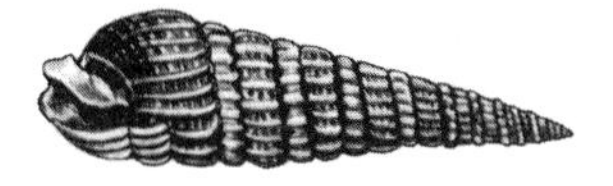

AUGER SHELL. *Terebra* spp.—feeds on small polychaetes and acorn worms. It has a poison gland associated with the proboscis to paralyze the prey.

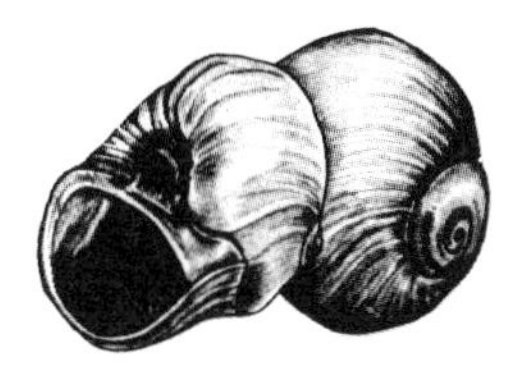

MOON-SHELL or SHARK-EYE SHELL. *Polinices duplicata*—is a snail with a radula (See Ocean Beach for description of radula) for drilling small, neat holes into the shells of bivalve molluscs. The egg case of this species, called a "sand collar," can often be found on the flats.

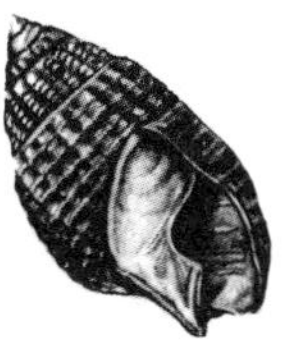

MUD SNAIL or SMOOTH BASKET SHELL. *Ilyanassa obsoleta*—is one of the most noticeable inhabitants of the flats. It crawls over the entire surface and may cluster in groups of several hundred in tidal pools. The mud snail is most common in areas where the bottom is slightly muddy. Its feeding habits are like those of the mottled dog whelk, *Nassarius*.

Bivalves filter water to obtain plankton and detritus. Some draw water and food into the shell through siphons; others simply open the shells so that water flows through. Most have an internal mucous "net" device for trapping organic particles.

PRICKLY COCKLE. *Trachycardium egmontianum*—extends siphons to the surface above its burrow in the sand, pulling in suspended food particles to a mucous net.

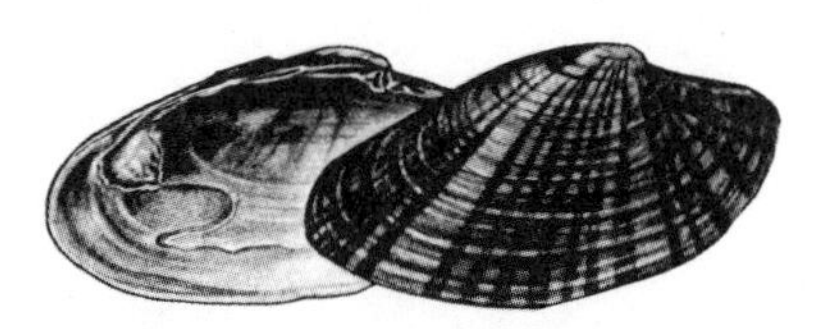

VENUS SUNRAY CLAM. *Macrocallista nimbosa*—burrows into the substrate. Siphons extended to the surface bring in currents of water and food particles to its mucous net.

EASTERN OYSTER. *Crassostrea virginica*—opens its shell at high tide to filter plankton. It has the ability to eject sand and mud taken in with its food. The razor sharp shells attach to any solid object, including other oysters in the lower intertidal zone. Once attached, the oyster remains immobile.

ATLANTIC BAY SCALLOP. *Argopecten irradians*—has ridges running from the top of the shell to the outer margins. The scallop is most common in eelgrass, where the young attach to the blades until large enough to survive on the sandy substrate. It feeds by opening the shell and filtering plankton.

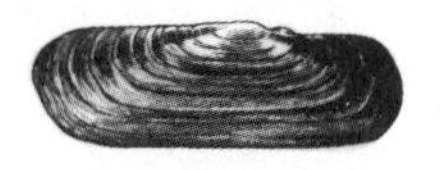

FALSE RAZOR CLAM. *Tagelus* spp.—may burrow to more than 30 cm below the surface, but moves upward for feeding. A long siphon draws in water and suspended food particles.

PEN SHELL. *Atrina serrata*—is a large, burrowing bivalve. The tips of the shells are left exposed at the surface, then open up to feed on suspended particles.

DISK SHELL. *Dosinia discus*—has a smooth, white shell. Siphons bring in water and suspended organic matter to the animal burrowed in the sand.

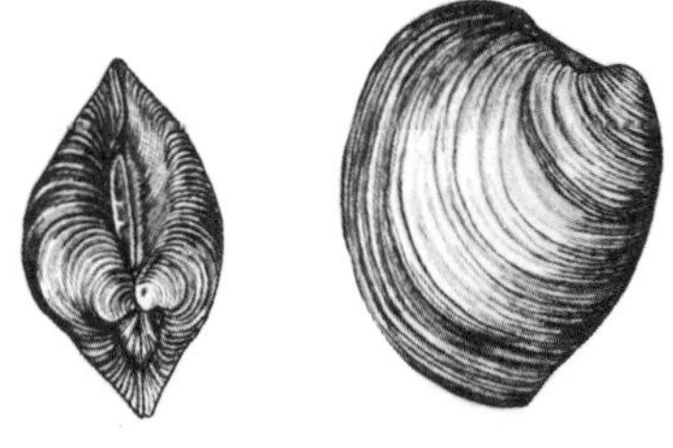

HARD CLAM. *Mercenaria mercenaria*—is the clam most commonly eaten by people in North Carolina. It extends short siphons into water above the sand surface to bring in plankton to a mucous net.

ARK SHELL. *Anadara ovalis*—brings in currents of water via siphons and feeds on plankton strained by the mucous net.

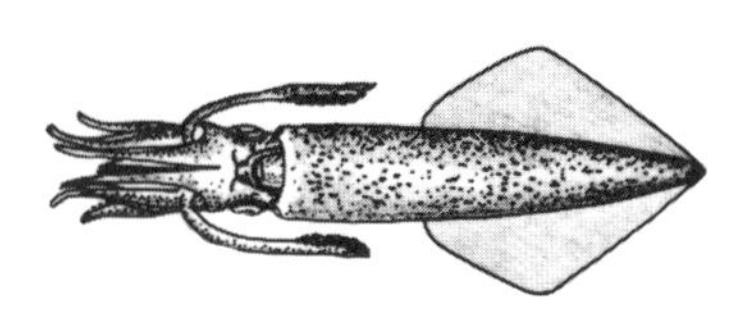

SQUID. *Loligo* spp.—is a mollusc closely related to the octopus. It generally remains in deeper channels of the sounds, but may move to flats at high tide. The squid is a rapid swimmer, capturing shrimps and fishes. It creates havoc with mackerel, although the fish may be larger, by moving through schools of the fish and biting off chunks of the prey with its sharp, beaklike mouth.

Annelid worms are common in this community. Many move freely through the sandy-muddy bottom and swim in water over the flats. Others build temporary tubes of sand in the substrate, or construct more elaborate permanent tubes whose "chimney top" openings dot the sand by the hundreds.

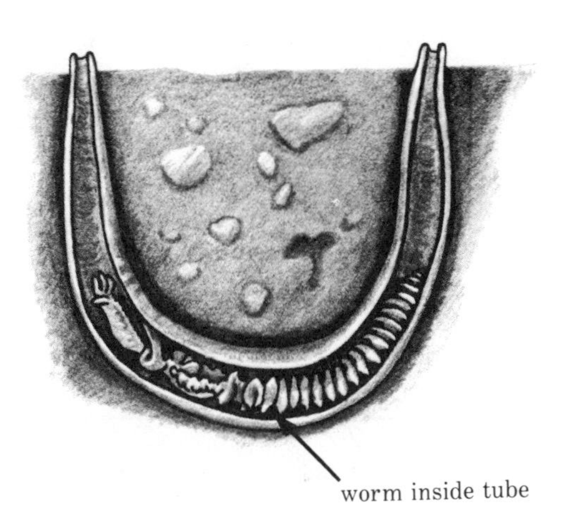

PARCHMENT TUBE WORM. *Chaetopterus* sp.—uses body secretions to build a permanent, U-shaped tube with a thick, leathery consistency. The tube has two chimney openings extended above the sand. At times the tubes are as dense as 8 or 9 per square meter. The worm is quite large, up to 30 cm long, and never leaves the tube. Currents of water are drawn into the tube by fanlike projections on the body. Food particles trapped in a mucous net on the body are passed to the mouth.

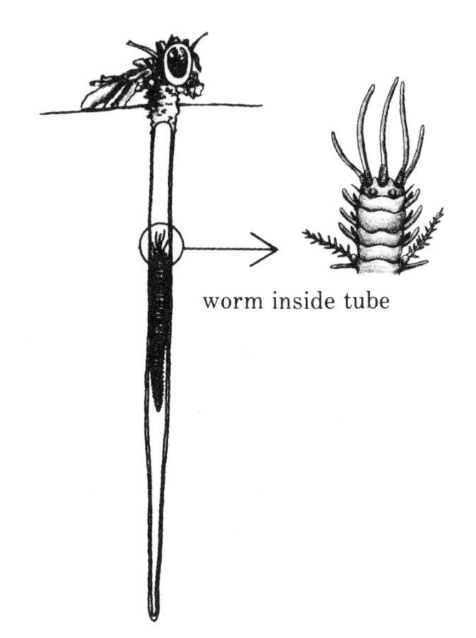

PLUMED WORM. *Diopatra* sp.—is a beautiful irridescent greenish red worm, abundant on mixed sand-mud bottoms. The permanent tube, which may extend 40 cm into the sand, has a single chimney with the top bent at a 90° angle. The worm often camouflages the top with seaweed and small shells. To feed, it extends its head and much of its body out of the tube opening and moves its proboscis in and out to catch tiny invertebrates.

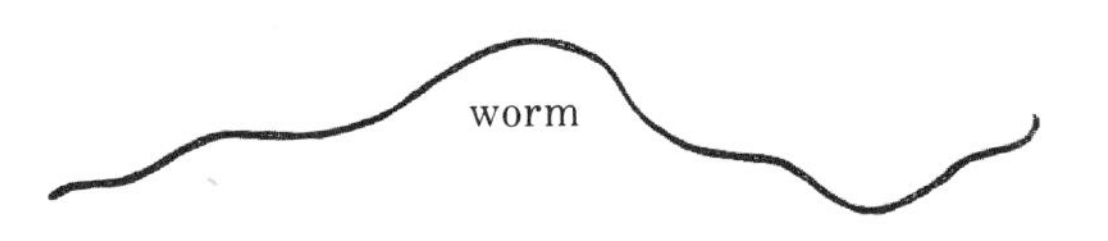

THREAD WORM. *Drilonereis longa*—resembles a long, red thread. The body is "elastic" and may stretch to 30 cm or more. This narrow, irridescent red worm moves freely through the subsurface sand and mud, feeding on organic detritus.

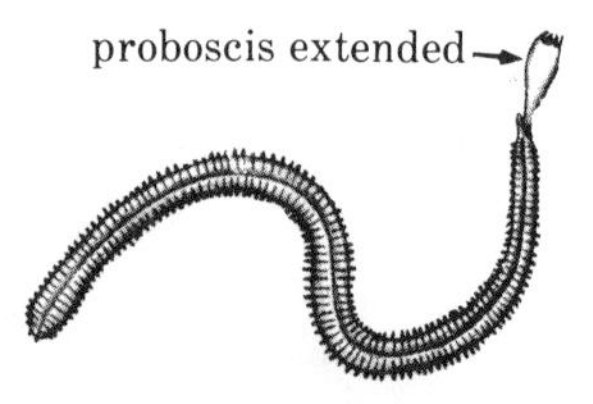

BEAK THROWER or BLOOD WORM. *Glycera* spp.—moves freely through sand or water by extending and contracting its beaklike head. Purple, pink, and red irridescent hues blend on its colorful body. The beak thrower is a food finder, moving about looking for other worms and small invertebrates and extending a long proboscis to capture prey.

LUGWORM. *Arenicola* sp.—excavates a burrow as deep as 30 cm beneath the surface. Coiled sediment castings deposited near the entrance to the burrow are evidence of its presence. The lugworm is a sediment ingester, moving through the substrate as deep as 15 to 30 cm and consuming sediment and the bacteria and organic detritus mixed with it.

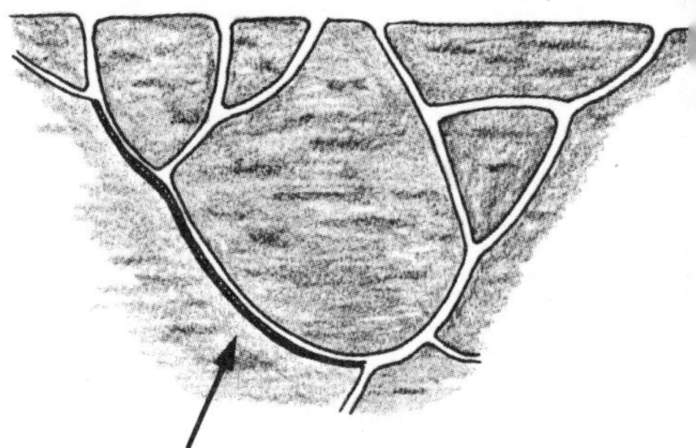

CLAM WORM. *Nereis* spp.—is among the most common and largest of coastal marine worms. Its brownish red body may reach 30 to 40 cm long. The worm burrows in sand or mud, making a temporary sand tube. It leaves the tube for voracious hunting of other worms and invertebrates, swimming through the water very rapidly in an undulating motion.

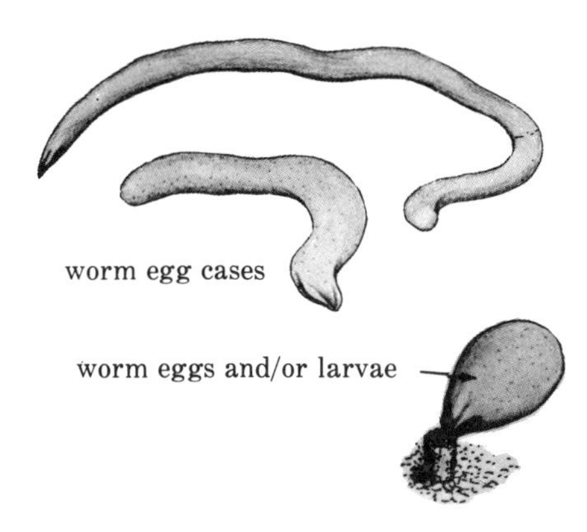

During summer, flats are marked with egg cases of such worms. These cases are gelatinous masses, long and narrow or short and rounded, usually attached to some solid object or to the sand. They commonly are mistaken for some sort of jellyfish since they are jellylike in substance, but a close look reveals thousands of minute red or white "specks" in the protective jelly casing. These are worm eggs and larvae.

Many arthropods find an ideal habitat on tidal flats, where the substrate is stable and food plentiful. The food web provides a variety of feeding niches for many different species. Shrimp-like crustaceans are common, actively swimming or hiding among blades of eelgrass. Some appear so similar that detailed examination is required to determine their exact identity.

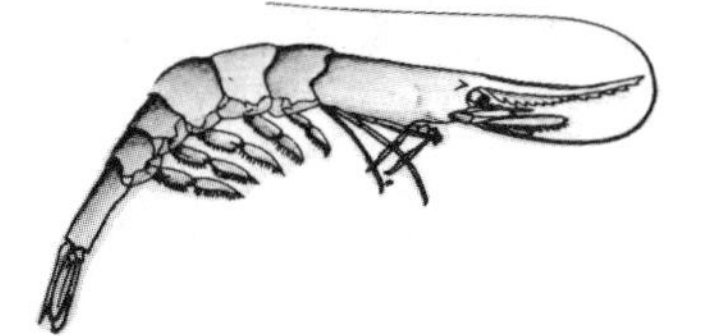

ARROW SHRIMP. *Tozeuma carolinensis*—is a small, thin animal, usually about 3 cm long, which commonly lives in eelgrass. It is most easily distinguished by the long, slender rostrum with 8 to 10 "teeth" on the bottom. The abdomen bends sharply, almost perpendicular to the body. The shrimp is bright or dark green, sometimes brownish or reddish, depending on background color.

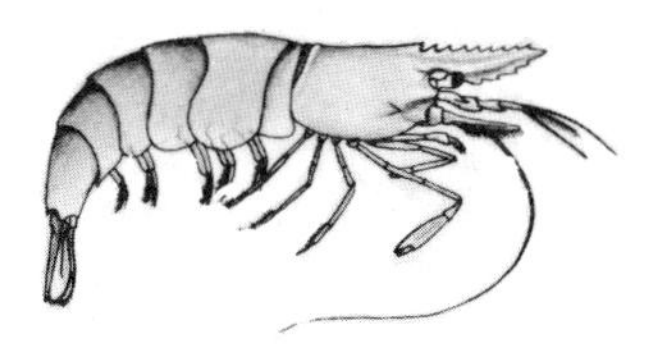

COMMON PRAWN. *Palaemonetes* spp.—has a small, translucent shrimplike form and is dotted with brownish spots. It often is mistaken for a young *Penaeus*, but seldom grows more than 5 cm in length. The rostrum, which tips slightly upward, appears a bit concave, with 8 to 9 "teeth" on top and 3 to 5 below. This prawn is common in eelgrass.

SHRIMP. *Penaeus* spp.—adults may be 13 to 15 cm long. Because of their size and taste, these shrimps are commonly eaten by people in North Carolina. Adult shrimps prey on fishes, molluscs, and other crustaceans and scavenge on dead bodies of such prey as well. Immature shrimps feed on detritus. All three types of *Penaeus* common on the North Carolina coast are named in the Ocean Beach section.

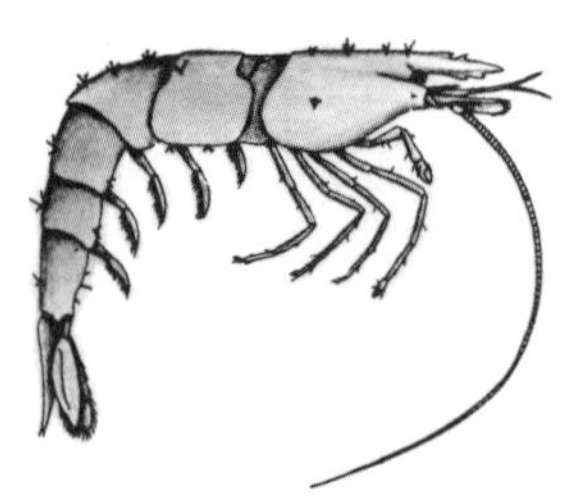

AESOP PRAWN. *Hippolyte* spp.—appears similar to the arrow shrimp, but has a shorter rostrum with 1 to 3 "teeth" on top and 1 to 3 on the bottom near the tip. The Aesop prawn reaches only 1.7 cm in length. The abdomen bends sharply downward, and the color varies from mottled brown or red to bright green, depending on background color. The prawn reportedly scavenges small organic tidbits.

Other shrimplike animals are much more elusive on the tidal flat since they live in burrows excavated in the substrate. Many holes in the sand are the openings to their burrows. Rapid digging often fails to uncover these animals, since they retreat to the bottom of the burrow 30 cm or more beneath the surface.

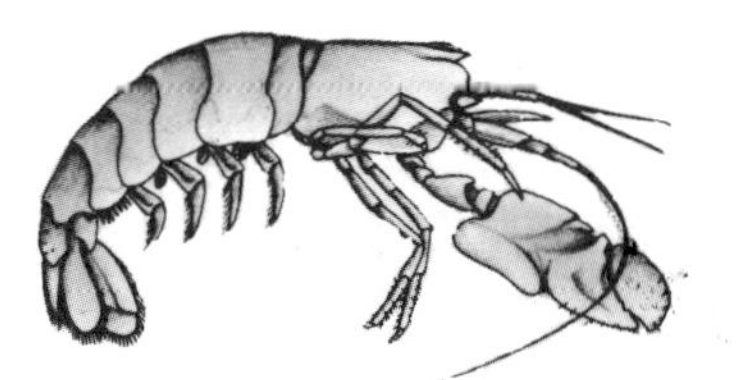

SNAPPING SHRIMP. *Alpheus* spp.—can be recognized by the loud snapping sounds it makes when disturbed in hiding among broken shells or stones and in its burrow. The first pair of legs is slender, but one leg ends in a large, outsized "claw." The "wrist" of this leg produces the sound. This shrimp grows to 5 cm long, has a minute rostrum, and has many translucent hues of color. The animal's exact feeding habits are unknown, but it probably feeds on small invertebrates.

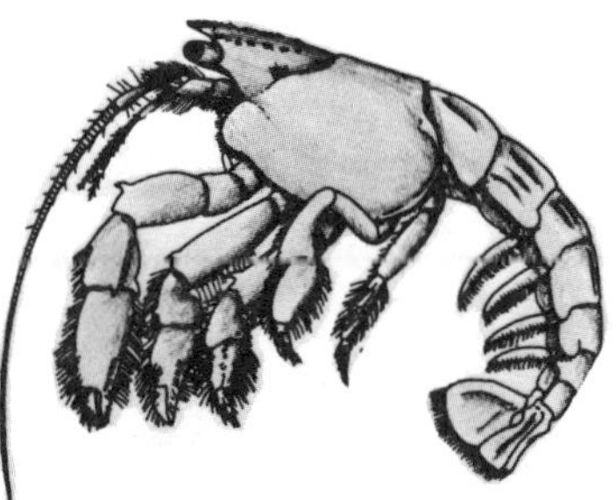

GHOST SHRIMP. *Callianassa* spp.—is not a true shrimp. It builds burrows, with many tunnels, as deep as 1 meter. Swimmerettes pull a current of water into the burrow where the animal strains plankton and detritus through hairy mouth parts. A ghost shrimp is about the same size as a mud shrimp, but the rostrum is minute and one of the front claws of the male is larger and longer than the other. The animal's common name derives from its pale "ghostly" color and the fact that it is never seen outside its burrow.

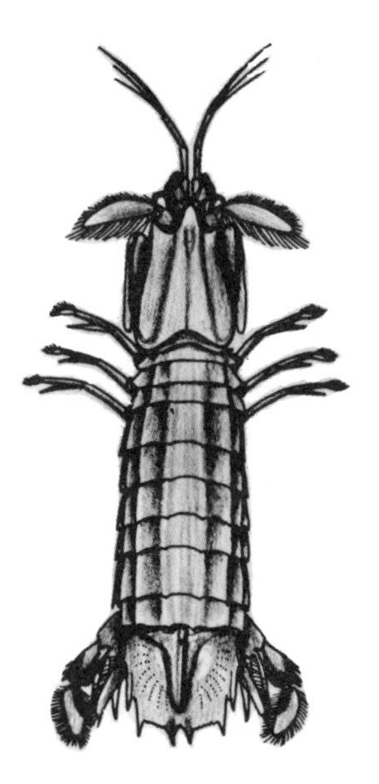

MANTIS SHRIMP. *Squilla empusa*—is a member of a crustacean group separate from shrimps and crabs. It resembles a praying mantis, with very long, slender front legs and eyes situated on long, narrow stalks. The body is broad and flattened, with translucent pale yellowish green hues. It digs burrows in muddy sand near and below the low tide line. The scythelike claws can inflict a sharp cut, so handlers should beware. This animal waits at its burrow entrance to capture prey or swims in pursuit of small fishes, crustaceans and other invertebrates.

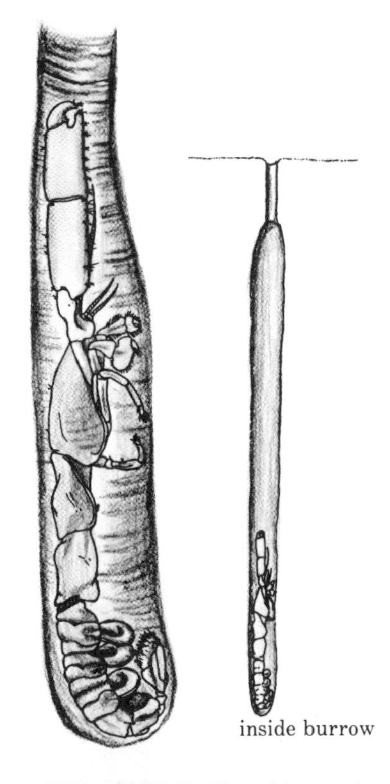

MUD SHRIMP. *Upogebia* sp.—is a shrimplike crustacean, more closely related to crabs and lobsters than to true shrimp. Common in muddy areas, its elaborate burrow may be 30 cm in depth and is usually near or below the low-tide line. At the entrance, the animal creates a current of water with its swimmerettes and strains the water for plankton and detritus. Tiny hairs covering the first two pairs of legs aid in filtering. A mud shrimp grows to about 5 to 6 cm long. It has a relatively large rostrum, a long, nearly flat carapace, and an abdomen that is held curved inward. The twin front clawed legs are the same size.

Crabs of all sizes and types frequent the flats, exhibiting an interesting variety of habits. Many crabs are rapid swimmers or are able to dig quickly into the sandy bottom. Others have made unique adjustments to life on the flats, living in or on the shells and bodies of other animals.

SPIDER CRAB. *Libinia* spp.—has only tiny claws which cannot pinch effectively. The crab seeks shelter in eelgrass. It is a deposit feeder, picking up fine organic particles with long, narrow front claws and passing the food to its mouth. The spider crab occasionally feeds on dead fishes and shrimps.

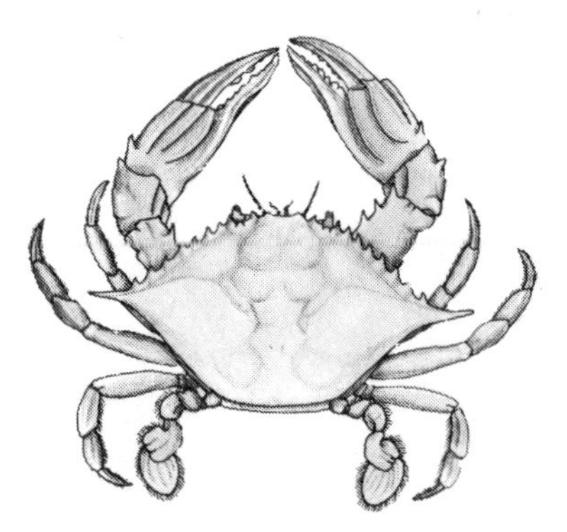

BLUE CRAB. *Callinectes sapidus* —is the common edible crab of coastal waters.It is able to inflict a painful wound with its two large claws. The blue crab scavenges on most dead animals, but also swims rapidly, catching shrimps and small fishes. The young are most often found hiding among eelgrass blades or digging rapidly under the sand for protective cover.

SAND FIDDLER CRAB. *Uca pugilator*—is easily observed at low tide, when hundreds can be seen scampering over the sand. The openings of their burrows pock the intertidal substrate. At low tide each crab reexcavates its burrow where it will retreat at high tide, and scurries about searching for food. A deposit feeder, the fiddler crab scoops up sand and mud and scrapes organic material from the grains with its mouth parts. The female has two small claws. The male has one small claw and a large claw resembling a "fiddle" which it waves about to defend its territory and to attract females.

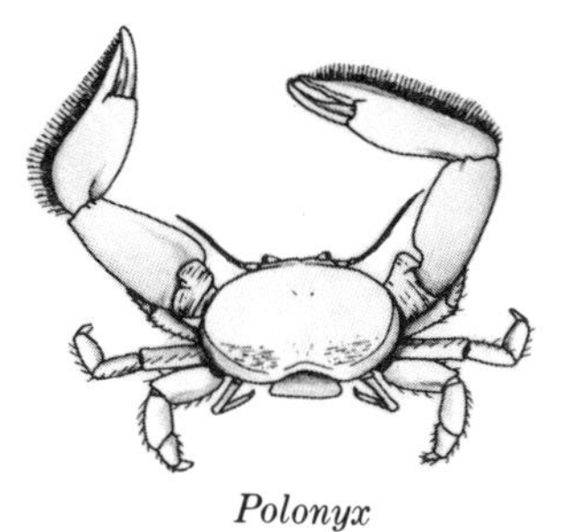

Polonyx

PEA CRABS. *Polyonyx* sp. and *Pinnixa* sp.—live in permanent tubes of worms such as the parchment tube worm. Here the tiny crabs, less than 1 cm long, are protected against predators and from being covered by sand. Pea crabs feed on planktonic food brought into the tube by the "host" worm.

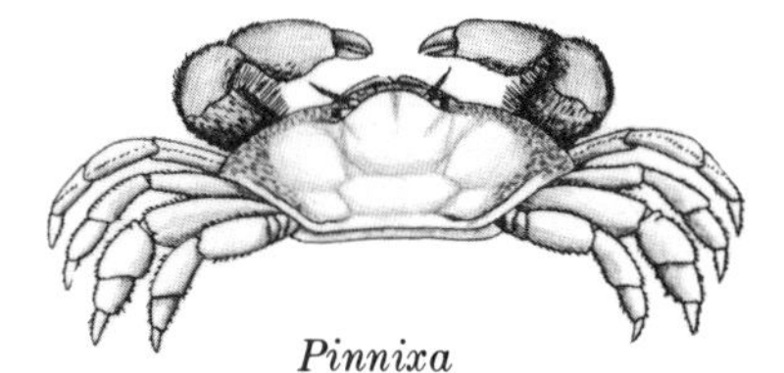

Pinnixa

OYSTER CRAB. *Pinnotheres ostreum*—less than 1 cm in size, this crab lives inside shells of the common oyster and other bivalves. When the bivalve opens its shells to feed, the crab filters water for detritus and plankton. It may at times eat the tissue of the "host" oyster.

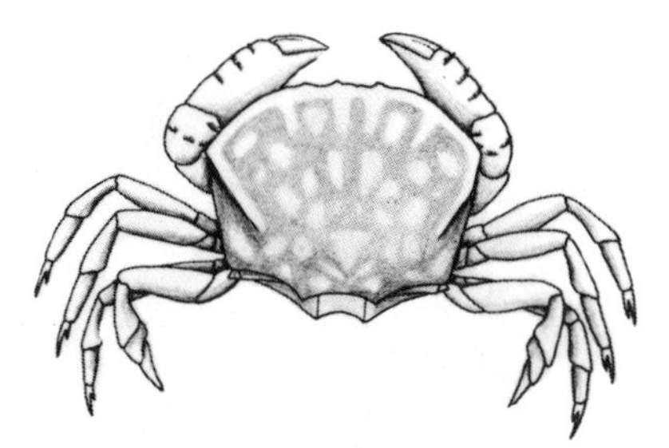

SAND DOLLAR CRAB. *Dissodactylus mellitae*—clings on the undersurface of sand dollars where it finds protection. It is tiny, less than 0.4 cm in size. The crab feeds on organic detritus located on its "host," or on detritus in sand through which the sand dollar moves.

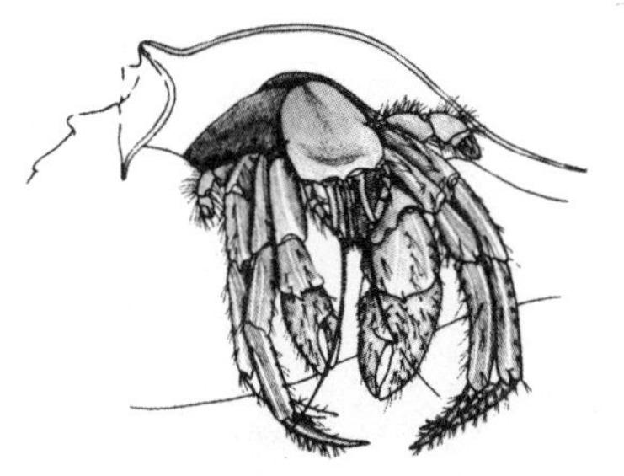

STRIPED HERMIT CRAB. *Clibanarius vittatus*—finds a home in abandoned shells of gastropods. The rear part of a hermit crab is elongated and curves downward, fitting into the spiral shape of the shells. This crab seldom leaves its shell except to find a larger one as it grows. Hermits are scavengers but also prey on small invertebrates. The hermit crabs, *Pagarus* spp., also are common on the flats.

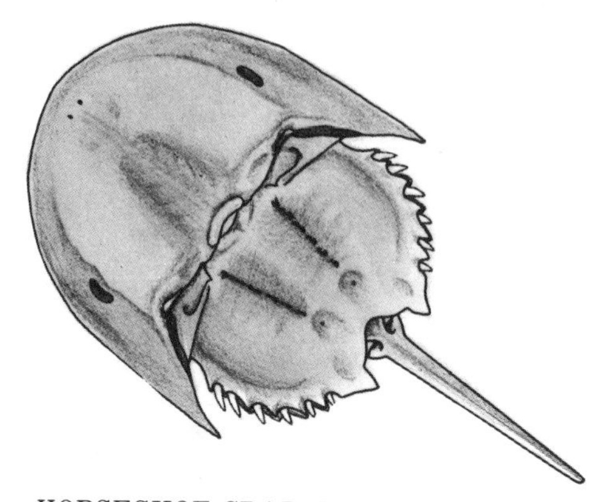

HORSESHOE CRAB. *Xiphosura polyphemus*—is not a crab at all, but an arachnid, the same class of arthropods that includes spiders and scorpions. Virtually unchanged for thousands of years, the species is considered a "living fossil" of the seashore. This distinctive animal moves slowly, its long "tail" extended behind. It is a food finder and crawls through sand and mud consuming worms, other small invertebrates, and algae.

One of the most misidentified animals of the tidal flat is the comb jelly. Often thought to be a jellyfish, since its body is a rounded glob of clear jelly, it is a member of the Ctenophora, an entirely different phylum from that of the jellyfish.

COMB JELLY. *Mnemiopsis* sp.—has a transparent, almost invisible, pear-shaped body, 7 to 10 cm long. Eight separate tiny ridges traverse the length of the body. Called a comb plate, each ridge has tiny, hairy projections like the teeth on a comb. Comb plates, iridescent during the day and fluorescent at night, are used to propel the animal through the water as it floats. It often occurs in dense groups, called "swarms," particularly during summer months. The comb jelly feeds on plankton.

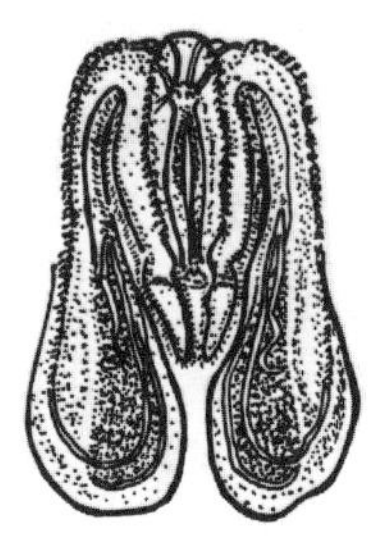

Urchins, sea cucumbers, and sand dollars are common echinoderms living in or on the sandy bottom.

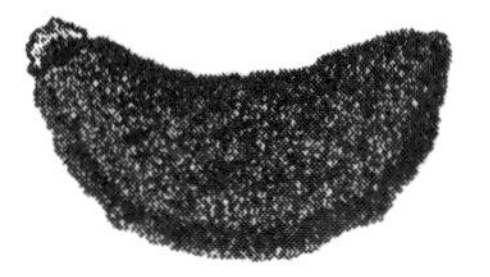

COMMON THYONE. *Thyone* sp.—looks like a dark cucumber. Slightly oval, the body is black, olive, or brown, and generally 10 to 13 cm long. The body is covered with small tube feet by which the animal moves. *Thyone* burrows into the substrate, keeping both ends above the sand surface and the middle portion buried. It is a deposit feeder, extending mucous-coated tentacles above the surface to trap organic sediment. The tentacles are then pulled into the mouth where food particles are scraped off.

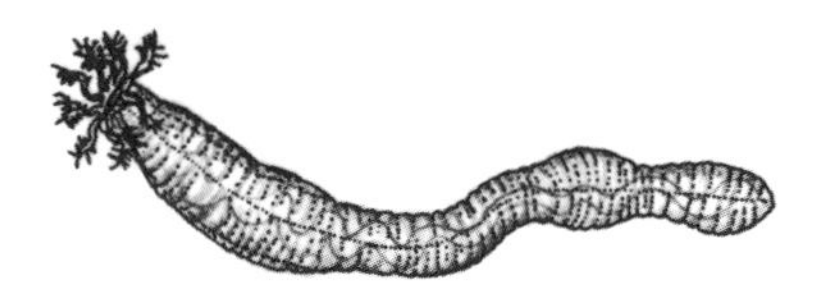

COMMON SYNAPTA. *Leptosynapta inhaerens*—is a sea cucumber which, unlike *Thyone*, has a white, transparent wormlike body. It, too, buries in sand, and grasps sediments with its tentacles, passing them to the mouth. Particles of sediment, sand, mud, and organic material can be observed passing through its body. *Leptosynapta* is much smaller than *Thyone*, being about 2.5 cm long.

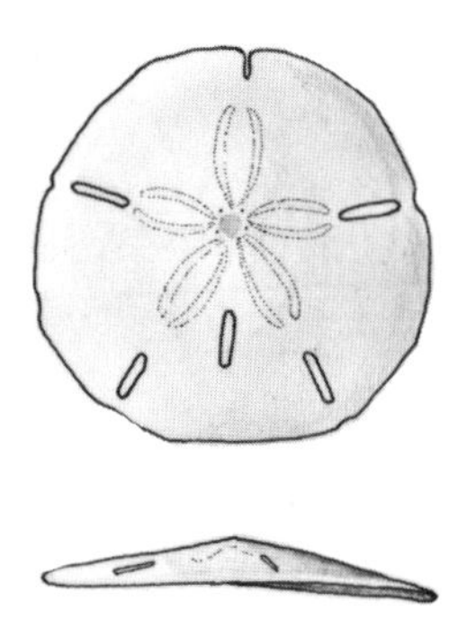

KEYHOLE URCHIN. *Mellita* spp.—moves through the sandy bottom. Minute spines and cilia on its body carry small organic food particles from the sand to the mouth on the underside.

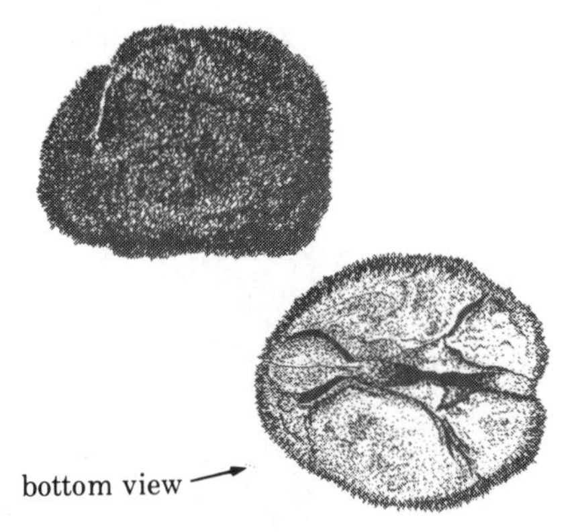

HEART URCHIN. *Moira atropos*—excavates a burrow that opens to the sand surface, secreting a cementlike substance that holds the burrow walls intact. This urchin is oval, colored a blend of yellow, red, and brown, and is generally 7.0 to 10 cm across. Its body is covered with sparse, short, flexible spines interspersed with short tentacles. The tentacles and spines capture and hold small animals and food particles and pass them along to the mouth on the underside.

The following bryozoans, sponges, and coelenterates are common on tidal flats. They are sessile and need a solid object to which they can attach.

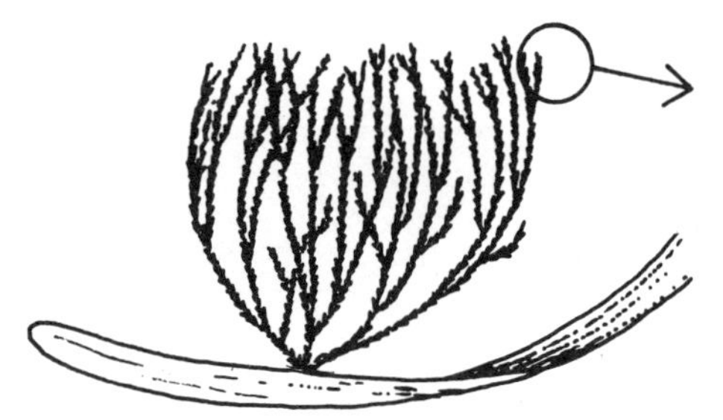

colony attached to eelgrass

individual home

MOSS BRYOZOAN. *Bugula* spp.—looks more like a plant than like an animal. It is called a "moss animal" because its appearance is similar to that of a dark brown to purple moss. The bryozoan occurs in thick, colonial clusters. Each individual in the colony has a crown of tentacles with cilia to aid in gathering plankton.

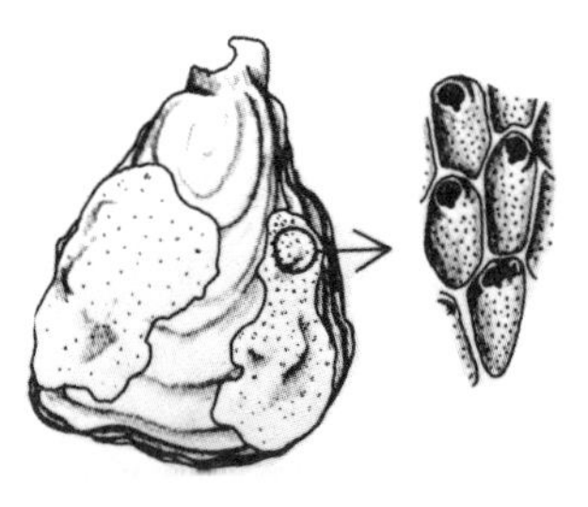

ENCRUSTING BRYOZOAN. *Schizoporella* sp.—unlike the other two bryozoans, grows as a flat crust, coating the surfaces of shells and rocks. It is a golden orange color. A close look reveals tiny openings in the "crust" through which tentacles extend to feed on plankton.

CORK-SCREW BRYOZOAN. *Amathia convoluta*—also is a colonial "moss animal" with many branches, each resembling a tiny, cream colored corkscrew. Each individual feeds on plankton by means of tiny tentacles.

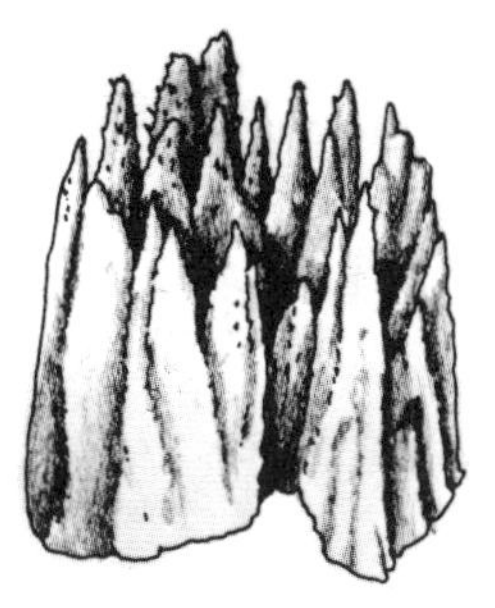

SUN SPONGE. *Hymeniacidon* sp.—also is a plankton feeder and colors the habitat with splashes of dull orange color, particularly during summer. The colony is a spongy mass, 5.0 to 10 cm high, with numerous small lobes 1.0 to 2.0 cm long projecting up. It settles densely on solid surfaces.

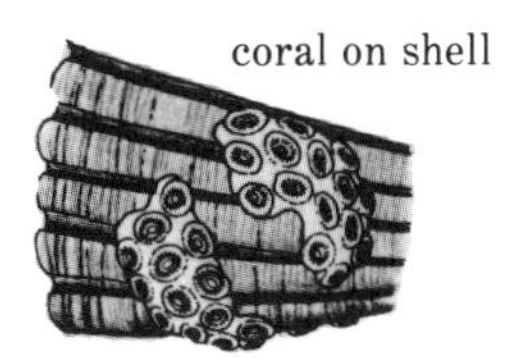

STAR CORAL. *Astrangia* spp.—is a whitish stony coral which grows in encrusting colonies. Each individual secretes a structure of calcium carbonate which joins with the structure of the adjacent individual and hardens. At the center of each is a starshaped elevation in which the live coral dwells. Each polyp, armed with stinging cells, extends to capture plankton and other small invertebrates.

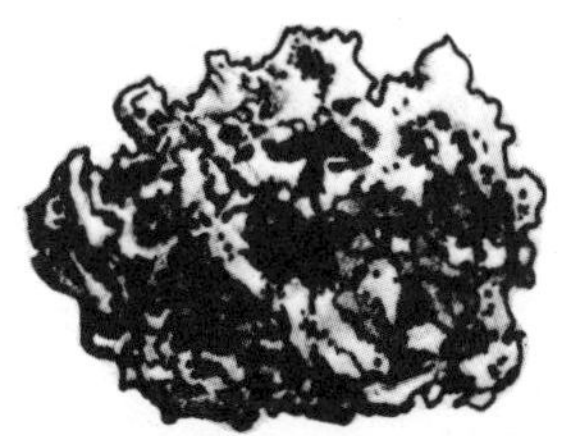

GARLIC SPONGE. *Lissodendoryx* sp.—is slate colored and has the strong aroma of garlic. Several overlapping lobes form an irregularly rounded colonial mass, usually 10 to 20 cm in size, with the consistency of a typical "bath sponge." The garlic sponge is a suspension feeder, filtering water through the colony to extract planktonic food. Many small commensal animals live among its lobes.

SOFT CORAL. *Leptogorgia virgulata*—exists as a colony, formed by many individual coral polyps. The colony, which is of a soft, flexible material, branches like a small tree. On the branches hundreds of polyps extend to feed on plankton. The color may be white, orange, rust, or yellow; but each colony is a single color.

Chordates like the following are among the highest forms of animal life in the sea, although from their outward appearance most do not seem particularly complex or advanced. Sea squirts, sea grapes, and sea pork are all suspension feeders, extracting plankton from water pulled through the body.

SEA SQUIRT. *Styela* spp.—its colony is up to 7.0 cm long, light brownish gray, and somewhat opaque. It has the feel of plastic and often has tiny knobs and ridges. Two or more "openings" or siphons on the body are used for circulating a current of water inside, from which plankton are extracted.

SEA GRAPE. *Molgula* spp.—looks like a small, opaque grape, with a grayish color. Often the insides can be seen through the semi-translucent body.

Although acorn worms look like worms, they actually are hemichordates.

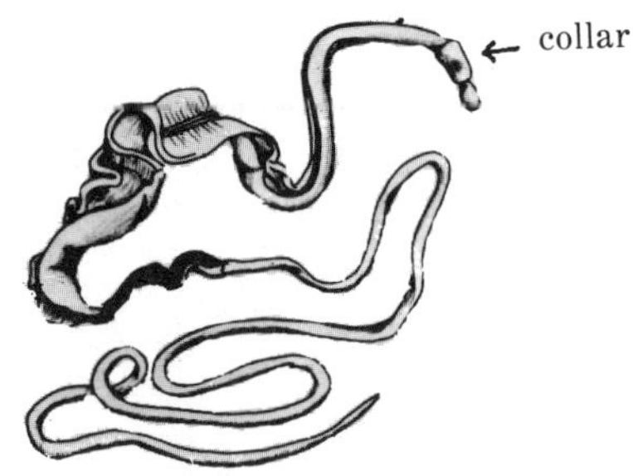

ACORN WORM. *Balanoglossus* spp.—can withdraw its proboscis into a collar located behind the head. The animal moves freely through the substrate and uses its proboscis to ingest muddy sediments which then pass through the body. Organic particles are extracted, and the remaining materials are discharged as a pile or "casting" on the surface. Such castings are a visible clue to the presence of this elusive animal.

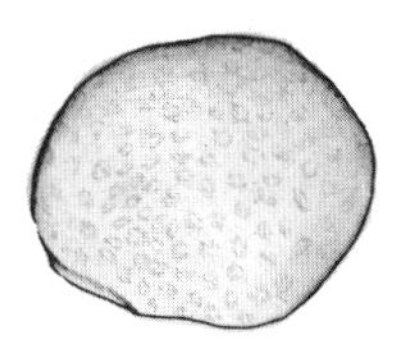

SEA PORK. *Amaroucium* sp.—is creamy peach colored and often mistaken for a piece of hard salt pork. Each colony, globular and like tough plastic, grows 8.0 to 10 cm across. Tiny red pinpoints mark individual members.

Many fishes become part of the tidal flat community at high tide, moving in to feed on the wealth of food available. Small fishes predominate, taking advantage of the protective cover of eelgrass. The young stages of larger fishes such as immature pinfish, spot, croaker, bluefish, and others also hide and feed among the thick blades of vegetation. (See other communities for illustrations of these.)

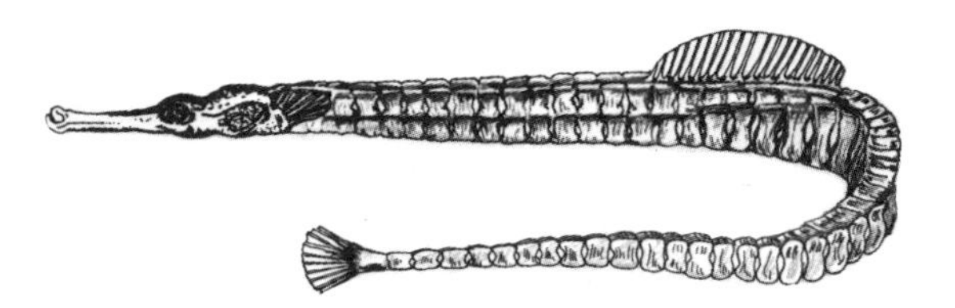

PIPEFISH. *Sygnathus* spp.—has a slender, snakelike body, 10 to 15 cm long, which blends perfectly with blades of eelgrass. It often assumes a vertical stance like the grass blades. The pipefish consumes plankton, and the long, tubular snout is well constructed for sucking in minute organisms.

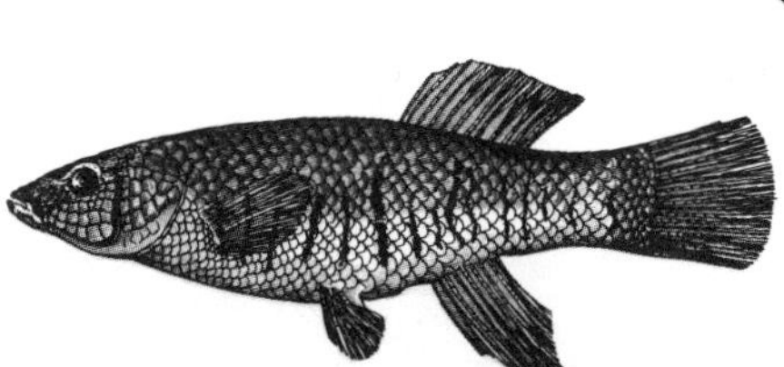

KILLIFISH. *Fundulus* spp.—generally reaches no more than 15 cm in length. It is very abundant on the flats and preys on a variety of small invertebrates.

FILEFISH. *Monocanthus hispidus*——has a small mouth, requiring it to feed on small invertebrates.

As the tide ebbs many birds move in to feed on animals stranded in tidal pools. Egg masses and larval stages of many tidal flat animals also are available as food. Herons wade in shallow margins, stalking minnows and crustaceans. Gulls scavenge for dead or helples animals of any type. Terns find good hunting in the subtidal fringe of the flats because fishes cannot easily escape in the shallow water. Sandpipers of several species forage the moist exposed surfaces for small animals. (These birds are illustrated in discussions of other communities.) On the rising tide birds disperse to other nearby communities for additional feeding, or gather in flocks on shoals to preen and rest.

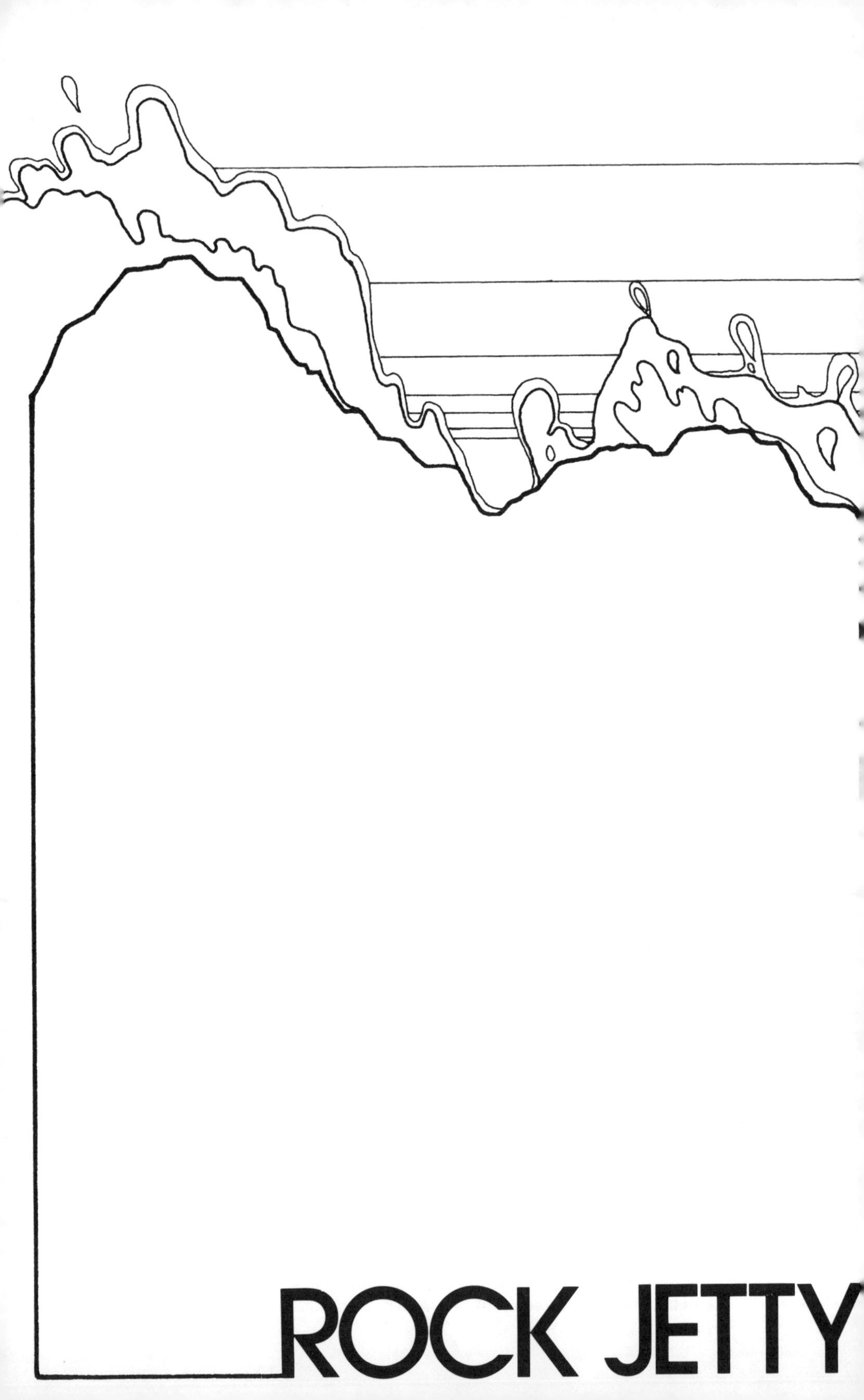
ROCK JETTY

Rock Jetty and Piling Habitat

The habitat

North Carolina lacks the solid shores characteristic of New England's rocky prominences and pebbly beaches. Marine plants and animals that attach themselves to solid objects must compete for suitable substrates on the mid-Atlantic coast, and encrusting organisms quickly cover the occasional shell remains and stones in the water. Solid formations are limited to man-made rock jetties and wood or cement pilings of docks, bridges, and seawalls located in both the ocean and the sounds.

Special features

—On jetties and pilings located where the water velocity is relatively great, waves strike and pound the solid surface. Such wave action may shred plants, and animals with delicate body structures may suffer.
—Strong currents carrying suspended sand may damage unprotected organisms, which may be pounded against the rocks, subjected to sand abrasion, or swept away.
—Dissolved oxygen is plentiful in water flowing rapidly over and through the habitat.
—The substrate consists primarily of steep vertical surfaces rather than the gradually sloping elevations of other habitats.
—Physical conditions change rapidly every six hours as tides move up and down the intertidal surfaces.
—Most species must directly face the forces of sun, wind, and water, since few are able to burrow into solid substrates. Wooden pilings, often riddled with the holes of burrowing animals, may eventually be destroyed.
—The solid substrate does not retain water when the tide ebbs, but occasionally water remains in depressions on rock jetties where small tidal pools may form.
—Following high tide, the upper intertidal zone in relatively calm water areas is exposed to air for prolonged periods. The chances of desiccation here are increased by sunlight striking the habitat and by winds blowing across already dry surfaces.
—The intertidal and supratidal substrates are exposed to extremes of heat and cold. Organisms must survive on sun-heated surfaces in summer and withstand freezing temperatures and even occasional ice abrasion in winter.
—Wave tossed water striking the intertidal habitat sends a spray up into the supratidal area. Certain plants and animals edge upward and survive in this area called the splash zone.

Adaptations of plants and animals

—Plants and animals on the steep surfaces of the jetty and piling settle in specific vertical zones, primarily on the basis of their ability to withstand varying periods of exposure to air. Plants are adapted to absorbing specific wavelengths of light which penetrate to specific depths in the water. These plants group densely together, and each species competes for the appropriate settling depth in the subtidal zone. The result is a spectacular display of vertical zonation of life throughout the habitat.

—Seaweeds and most animals here have special holdfast devices for attaching to the substrate.

—Attaching organisms include plants and animals that settle permanently and animals which "hold on" but are able to move slowly over the substrate. Such motile animals have special structures to aid them in clinging. Many have tube feet or other devices with a strong, suctionlike action.

—Many animals have heavy shells and sturdy bodies that prevent damage from waves and swirling sand. The body shape often is flattened or domelike to help reduce the effects of waves and pounding water. Other animals and plants have flexible bodies which allow them to move with the waves.

—Some animals live on undersides of rocks and in crevices, seeking protection from predators and waves and avoiding desiccation.

—Many animals living on pilings and jetties are protected from drying by a nonporous body covering. Others can tightly seal any opening in their shells by means of a strong "lid" called an operculum and a mucous seal.

—Some animals burrow into wood pilings for added protection. Many tiny animals live among the holdfast organs of hydroids and algae blanketing the substrates. Large numbers of minute crustaceans are protected in the thickly matted stalks of attaching hydroids.

—Certain blue-green algae secrete a slimy, gelatinous coating which prevents desiccation and enables them to live in the splash zone.

The community

Most of the plants and animals which make up the jetty and piling community have one thing in common—some special means of attaching or otherwise holding to the solid substrate. Competition for settling space is severe, and crowding becomes a pattern of life. The community contains many attached microscopic plants and animals, which along with larger seaweeds, support animals that graze surfaces and scrape off tiny organisms for food. Most animals in the community are suspension feeders and depend on plankton and detritus swept in from nearby sounds or the ocean. Therefore, most feeding activity takes place when the animals are covered by water rather than at low tide. Because little detritus settles on the substrate, sediment ingesters and deposit feeders are virtually absent. Despite the hazardous conditions of existence, jetty and piling communities thrive.

Typical organisms

Subtidal life

Plankton and other microscopic algae

Phytoplankton and zooplankton, both vitally important to the community, are bountiful in the surrounding water. Since little or no plankton settles on the substrate, animals must filter plankton from the water as it sweeps by in currents. (See Ocean Beach for examples of plankton.) An abundant film of microscopic algae coats the substrate and many animals depend on this algae for energy.

Larger plants and animals

Below the low tide line are sessile plants and animals that require water coverage at all times. Vertical zonation of plants and animals is noticeable but becomes progressively less pronounced at greater depths. Brown algae, anemones, bryozoans, sponges, sea squirts, and hydroids cluster just below the low tide line where they are exposed to air only at extremely low tides. Others, such as soft corals, most red algae, and white sea urchins, are not seen as easily, for they occupy the lowest depths. Different plant species may occur at different times in the community, varying in summer and winter.

Brown algae, such as the following, cluster densely at and below the low tide line.

SARGASSUM SEAWEED. *Sargassum* spp.—is represented by several attaching forms. *Sargassum natans* occurs in summer and *S. filipendula* grows the year around. Berrylike air bladders aid in flotation.

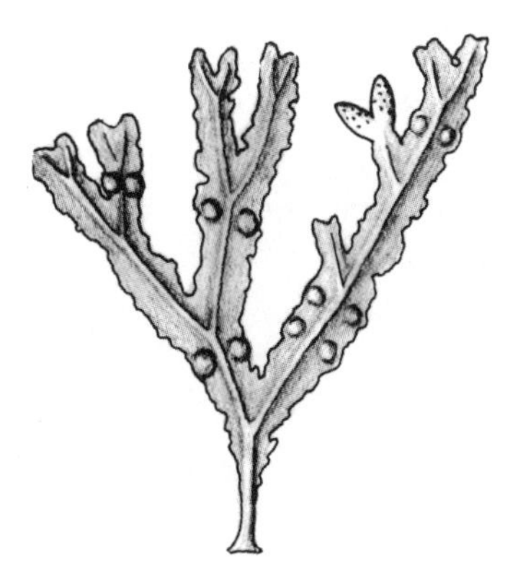

ROCKWEED. *Fucus* spp.—either brown or olive-green, rockweed is most abundant in spring. The blades, rather flat with many branches, have a thick, leathery texture. Numerous bulblike air bladders on the stems aid some species in floating.

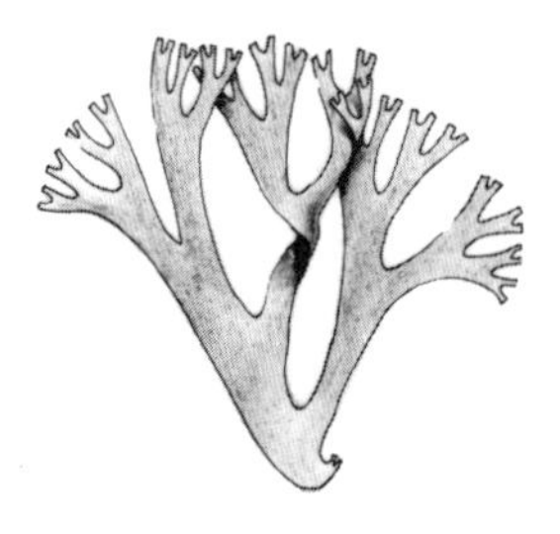

DICTYOTA. *Dictyota* spp.—is abundant in summer and early fall. Its fronds are like flat, olive brown membranes, about 6.0 mm wide, that fork into narrow branches, resulting in 15 to 25 cm-long tufts of growth.

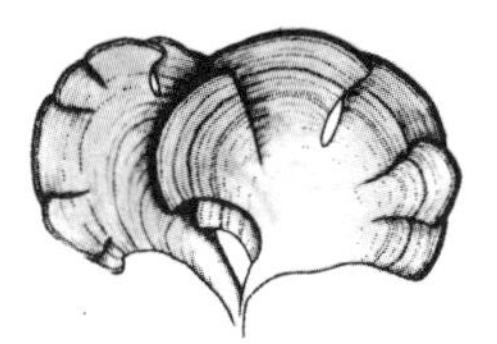

PEACOCK'S TAIL. *Padina vickersias*—resembles a miniature, light brown peacock's tail. The fronds are fan-shaped and rubbery, with dark brown, calcareous rings swirling around them. Peacock's tail flourishes in summer, but is not apparent in winter.

ECTOCARPUS. *Ectocarpus* spp.—consists of dense tawny plumes of loosely tangled filaments with many branches of hairlike fineness. The plant, 5 to 10 cm long, grows in profusion during winter and spring months on rock jetties, tidal flats, or wherever there is a suitable surface for attachment. It may also grow on top of other seaweeds. *Giffordia* spp., which appears virtually identical to *Ectocarpus*, replaces it in summer and fall.

Immediately below the zone of brown algae, such red algae as the following abound.

SEWING-THREAD SEAWEED. *Gracilaria* spp.—is purplish red to olive green and has many thin branches that do not tangle. Different species occur in spring and summer. The drifting phase, used commercially to make agar, is abundant also.

COULTER'S SEAWEED. *Neoagardhiella* spp.—most abundant in spring, it grows in dense clusters of 20 or more strands. Tapering at the ends, the pinkish white branches are smooth, round, and fleshy like spaghetti.

HYPNEA. *Hypnea* spp.—attains its greatest growth during summer. Although it is a red alga, gray-green pigments often dominate its color. The main stem's many irregularly spaced branches also branch, giving each plant a fragile, bushy appearance. The ends of the long branches are "naked" and bent like small hooks.

CHENILLE-WEED SEAWEED. *Dasya* spp.—most abundant in April, chenille-weed is usually gone by June. Its small, rounded, reddish purple branches are densely covered with a fine, chenillelike fringe.

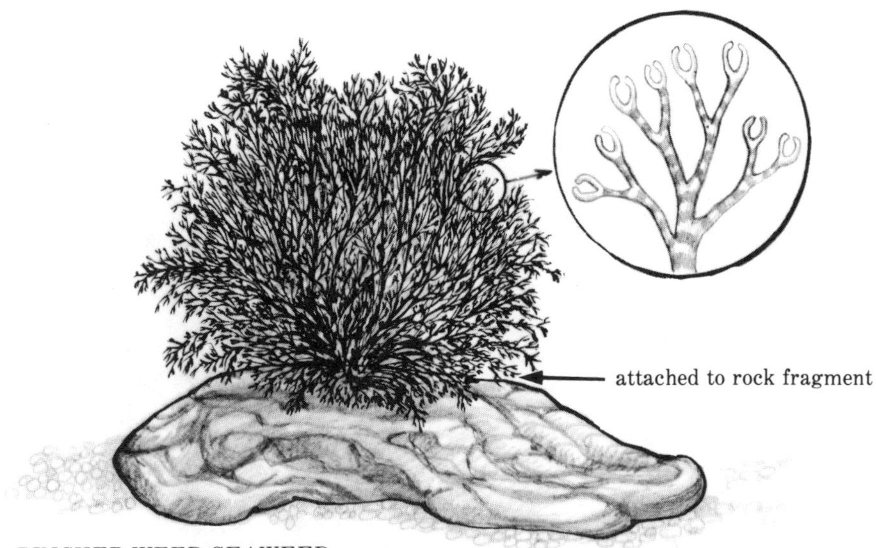

PINCHER-WEED SEAWEED. *Ceramium* spp.—abundant in spring, this delicate plant has branching reddish-purple filaments with forked ends that resemble tiny claws or pinchers.

Most of the following animals attach themselves permanently or move slowly over the subtidal substrate.

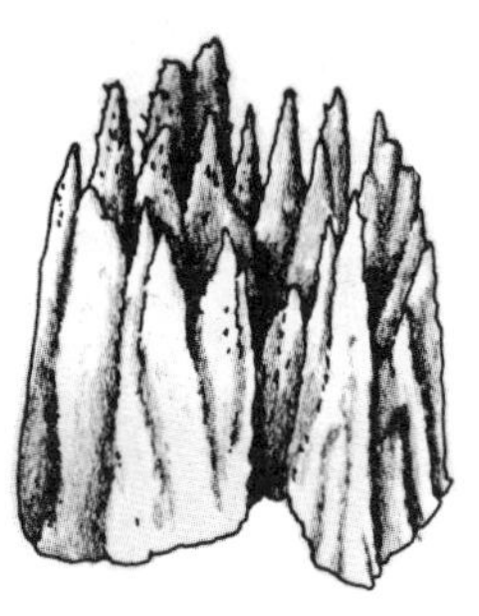

SUN SPONGE. *Hymeniacidon* sp.—distinguished by its dull orange color, this sponge is a typical mass 5.0 to 10 cm tall. Many small lobes extend from the colony's surface. The sun sponge also grows in the intertidal zone where numerous clumps provide noticeable color. A suspension feeder, the sponge extracts plankton from water it takes in via openings in the colony's surface.

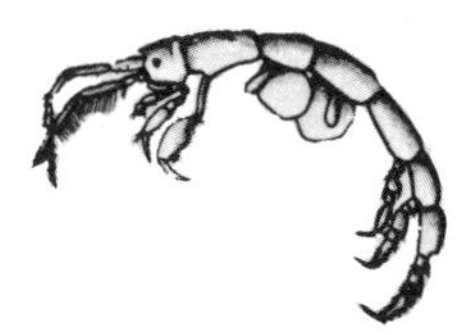

AMPHIPOD. *Caprella* spp.—a minute arthropod found scurrying among attached hydroids, moss bryozoans, and seaweeds. Amphipods often occur by the hundreds among these organisms, which not only constitute a food source but provide protective cover as well.

HYDROID. *Tubularia crocea*—is a relative of the sea anemone and occurs in both calm and wavetossed areas, growing in dense clusters that resemble tiny pink bouquets. Whitish colored "branches" are about 2 cm long, topped with two series of 20 to 40 pink tentacles which surround the mouth and capture plankton.

SEA ANEMONE. *Aiptasia pallida*—moves so slowly its progress is virtually undetectable. Its many top tentacles give it the appearance of a flower, while the base is a cuplike device that attaches tightly to the substrate. The sea anemone is a tentacle feeder, equipped with stinging cells to paralyze small prey.

COMMON SEA SQUIRT. *Styela* spp.—has an outer covering or tunic which feels like semihard plastic. Water is pumped into and through a central cavity in the body where suspended plankton are extracted.

RED BEARD SPONGE. *Microciona* sp.—is common wherever there is a solid surface to which it can attach. This sponge forms a colony with clusters of bright red orange, fingerlike lobes, up to 15 cm high. It filter feeds on plankton.

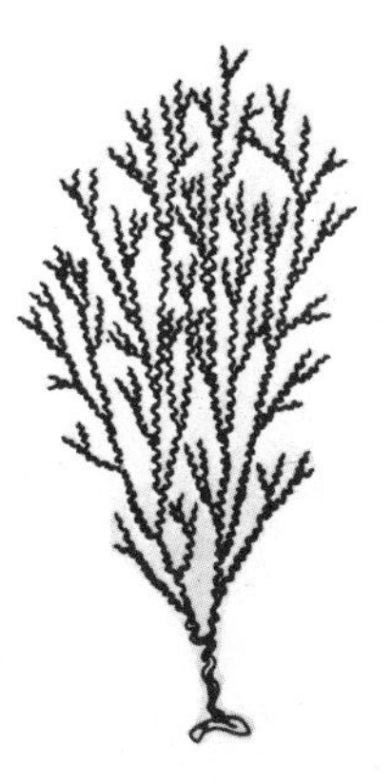

CORKSCREW BRYOZOAN. *Amathia convoluta*—is colonial, with many tentacles extending to filter plankton. It is a dark cream color with numerous stalks, each about 2.5 cm long, that spiral like a corkscrew.

YELLOW SULFUR SPONGE. *Aplysilla* sp.—is a flat, bright yellow sponge colony which encrusts solid objects. Easy to identify, it is a soft, slightly elastic mass with minute, sharply pointed cones rising from the surface. The sulfur sponge, like the sun sponge, is a filter feeder.

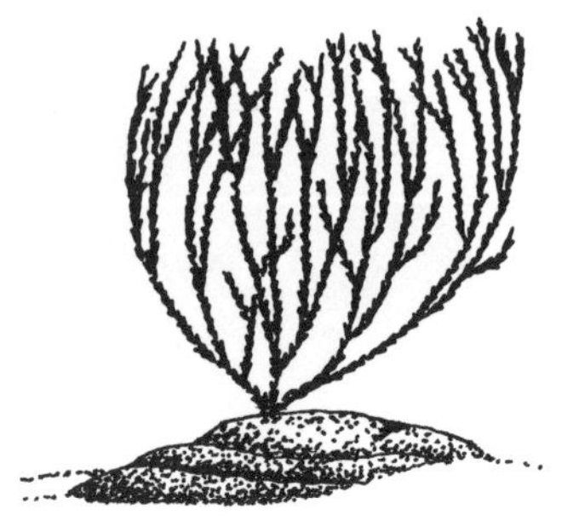

MOSS BRYOZOAN. *Bugula* spp.—is an abundant purplish brown animal that looks much like moss with many tiny branches. It is a colonial form with hundreds of minute individuals which extend tentacles from the "branches" to filter plankton.

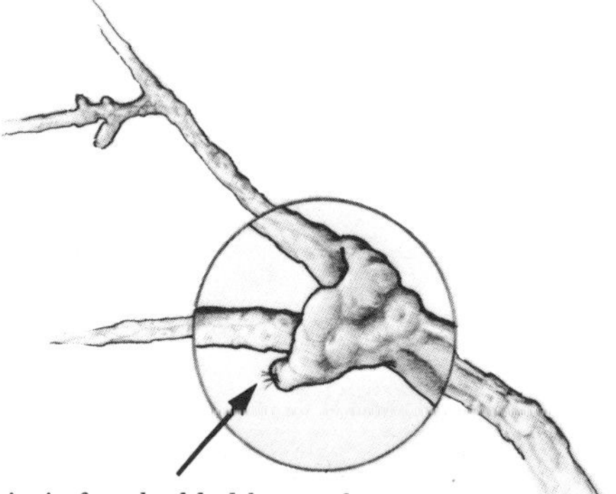

BARNACLE. *Balanus galeatus*—may embed itself while young in a soft coral colony. Eventually the coral completely encloses the barnacle except for a small opening through which the barnacle extends its cirri to feed on plankton.

WINGED OYSTER. *Pteria colymbus*—is a suspension feeder with an internal mucous net for straining plankton. The winged oyster often clings to branches of soft coral.

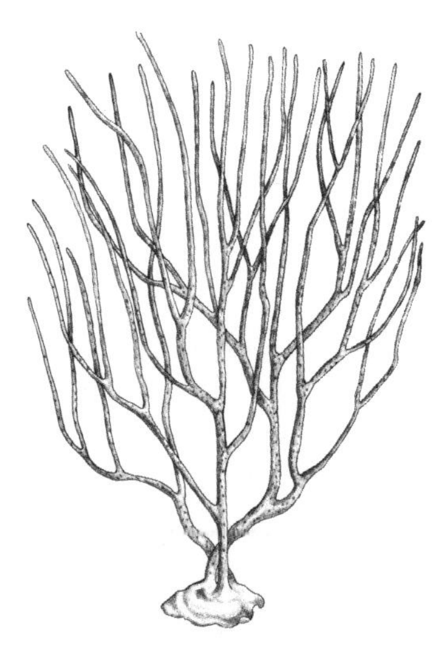

SOFT CORAL. *Leptogorgia virgulata*—grows as a colony consisting of a wirelike skeleton with several branches extending from a central "stalk". Because it is composed of a substance called gorgonin that is soft in comparison to the stony covering of other corals, the colony is flexible and sways with currents and waves. A strong holdfast at the base provides a firm attachment. Numerous polyps with feeding tentacles extend from the branches to capture suspended plankton. Each colony is a single color, but different ones may be orange, yellow, rust, or white.

PURPLE SEA URCHIN. *Arbacia punctulata*—like its relatives the starfish and brittle star, this urchin has hundreds of tube feet to hold to the substrate. These tube feet extend beyond the long, sharp, movable spines, enabling the urchin to creep over the substrate. It moves with the tides and hides in crevices to avoid light. Microscopic algae and animals clinging to the substrate provide its food. A special chewing apparatus called "Aristotle's Lantern", located on the underside, has fine, toothlike plates for scraping food.

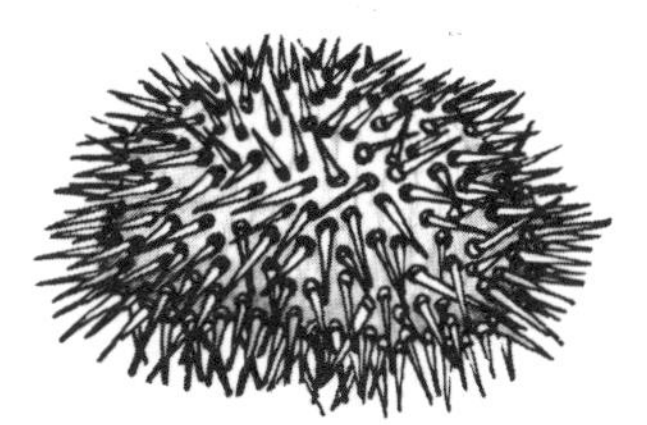

WHITE SEA URCHIN. *Lytechinus variegatus*—is creamy white and exhibits essentially the same life habits and form as the purple sea urchin but has shorter spines. The white sea urchin often stays at greater depths than does its relative, possibly avoiding direct competition.

While many types of fishes and other animals swim in waters flowing by jetties and pilings, most never become an integral part of the community. Their association with the community is fleeting and they do not participate in the food web. However, at least two kinds of fishes do belong to the community.

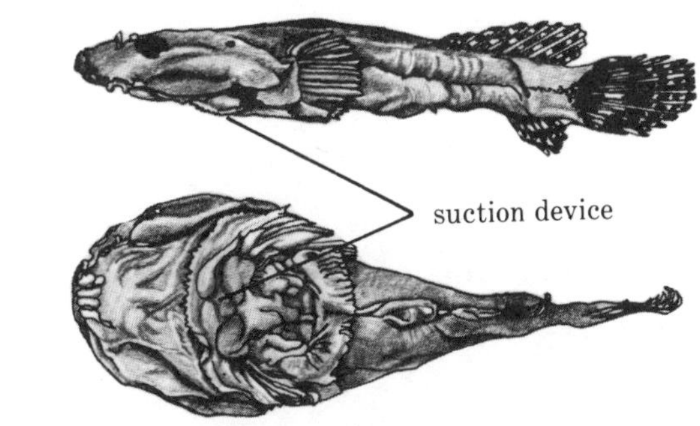

CLING-FISH. *Gobiesox* sp.—is small and uniquely adapted for life in this habitat. On its underside near the head is a cuplike suction device used to attach to solid substrates even in rough and swirling water. Thus a cling-fish can rest securely in and among rocks and crevices and still periodically venture out to feed on small crustaceans.

SHEEPHEAD. *Archosargus* sp.—is common and feeds on bottom dwelling shrimps and crabs, and on small molluscs and plants. It has sharp teeth to scrape and eat barnacles from the jetty and piling.

Intertidal life

Intertidal plant life is abundant. Green algae cluster densely together and compete for vertical space. A few red algae grow into the intertidal zone and blue-green algae extend even into the splash zone.

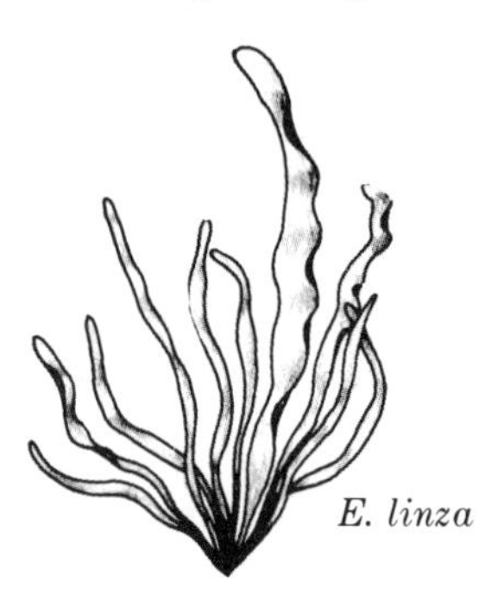

E. linza

SEA HAIR. *Enteromorpha* spp.—also called Mermaid's hair or Link confetti, it is represented by several different species in the middle to lower intertidal zone. Sea hair coats rocks and pilings in winter, coloring the substrate bright green. Other species, with only slight variations in appearance, take over in summer. Sea hair, although similar to sea lettuce with its green, paper thin blades, grows in clusters of threadlike or confettilike strands which may be from 2.5 to 30 cm long, depending on the species.

layer of alga growing on rock

BEAUTIFUL HAIR. *Calothrix* spp.—occurs high in the intertidal zone with another blue-green alga, *Lyngbya* spp. They form a fringe of fine tufts into the splash zone. Minute filaments 1 cm long are coated with a gelatinous sheath to protect against desiccation. Beautiful hair appears black since the filaments catch and retain muddy silt suspended in the water. Often it forms continuous slimy, spongy layers which can be scraped off rocks with a knife. Beware of slipping on this alga.

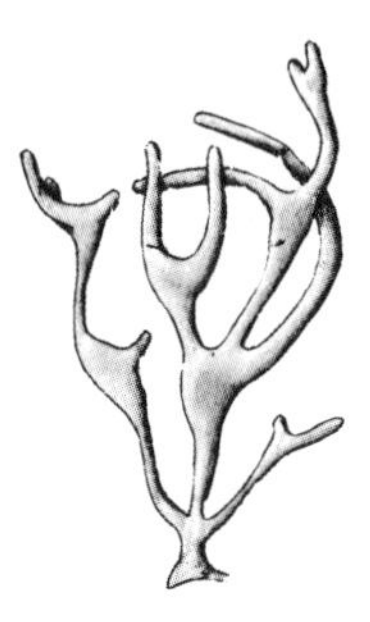

SPONGE SEAWEED. *Codium* spp.—is a green alga sometimes called "dead man's finger" seaweed. Feel one and you will understand why; its fingerlike dark green shoots, up to 10 cm long, feel spongy. The shoots fork at frequent intervals. When torn apart the broken ends will contract, sealing themselves immediately to avoid loss of water. This seaweed grows in the lower intertidal zone in summer and early fall.

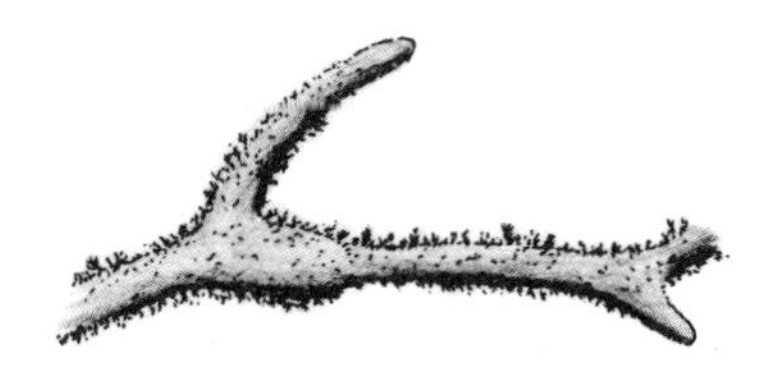

Acrochaetium growing on sponge seaweed

ACROCHAETIUM SEAWEED. *Acrochaetium* sp.—is a red alga which grows in dense wine-red clusters on top of other, larger algae, especially sponge seaweed. Acrochaetium is very delicate in appearance, with fine hairy filaments up to 1.5 cm long. It floats at high tide and forms a flat, silky mass as the tide recedes.

PURPLE-DYE SEAWEED. *Porphyra* spp.—also a red alga, this seaweed is a spring through summer resident. It looks much like sea lettuce, except for its purplish pink color. Commonly growing in the intertidal zone along with sea lettuce, it has paper thin blades with broadly wavy edges.

SEA LETTUCE. *Ulva fasciata*—has the same broad, flat, tissue thin blades as its cousin, *U. lactuca*, common to tidal flats, but is smaller in size. Its green to yellow green growth adds a splash of color to the higher intertidal zone, particularly in spring.

The intertidal zone is heavily encrusted with sessile animals which either can close their shells or otherwise seal themselves to avoid drying as the tide recedes.

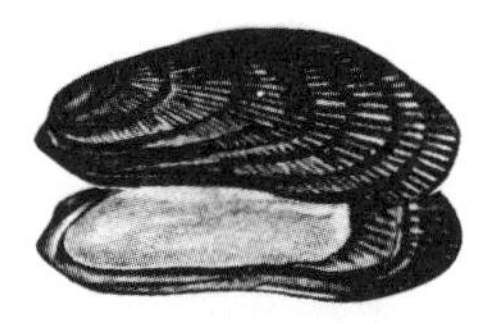

RIBBED MUSSEL. *Modiolus* sp.—locates below and occasionally among oysters on the substrate. It is a suspension feeder, with a molluscan-type mucous net for straining plankton. Numerous strong ridges or ribs run the length of the shell. Each mussel holds to the substrate by strong byssal threads secreted at the base of the shell.

EASTERN OYSTER. *Crassostrea virginica*—settles below the zone of barnacle growth and dominates the middle and lower intertidal area. It is a molluscan-type suspension feeder, opening its tightly closed shells at high tide to feed on plankton. In more exposed areas of the intertidal zone mussels may be more abundant than oysters, which are generally found where water velocity is moderate.

oyster drill

dye drill

ATLANTIC OYSTER DRILL *Urosalpinx cinerea*, and FLORIDA DYE SHELL. *Thais haemastoma*—are small but destructive snails which prey on oysters and barnacles. With a radula they drill a hole into bivalve molluscs, weakening the muscles so that the victim gapes open. (See Ocean Beach for explanation of a radula.) The adult dye shell is larger than the oyster drill, and both are food-finders.

ROCK BARNACLE. *Balanus* spp.—dominates the upper intertidal zone, and often is mistaken for a mollusc because of its hard, shell-like covering. The barnacle is a crustacean more closely related to crabs than to oysters. The calcareous plates ("shell") close at low tide to avoid desiccation and open at high tide so suspended plankton can be filtered from the water.

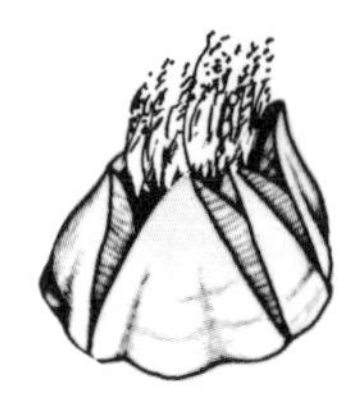

Motile animals, such as the following ones which need to be covered by water, move up and down with the tides to feed in the intertidal zone.

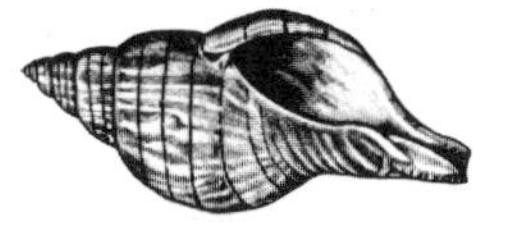

BANDED TULIP SHELL. *Fasciolaria hunteria*—is a large snail having a proboscis with a radula for boring into molluscan prey. When the banded tulip is exposed to air or endangered by predators, a thick operculum seals the animal in its shell. Thin, dark bands of coloration on the shell help identify this species.

SEA SPIDER. *Anoplodactylus lentus*—is a pycnogonid with long, slender walking legs which encourage the common name. However, it is not a spider. Its agile legs allow it to move about with the tides. Dark purple, the sea spider crawls over algae and hydroids on the jetty and pilings, feeding on anemones, on the polyps of hydroids and corals, and on bryozoans.

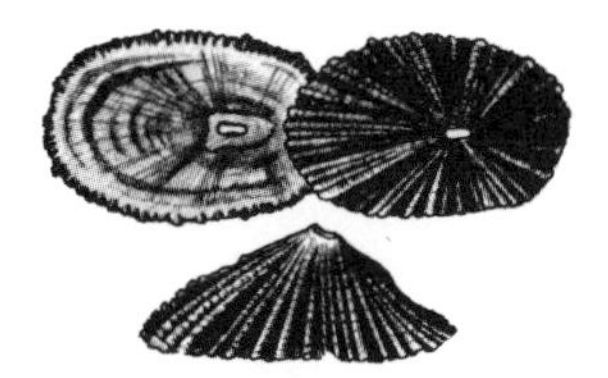

KEYHOLE LIMPET. *Diodora cayenensis*—moves on its large suctionlike "foot" into the intertidal zone during high tide to graze on microscopic algae. This species can survive exposure to air during the few hours of low tide. The limpet grinds away at the hard substrate (even rock) to hollow out a small depression into which it settles tightly, leaving at times to feed.

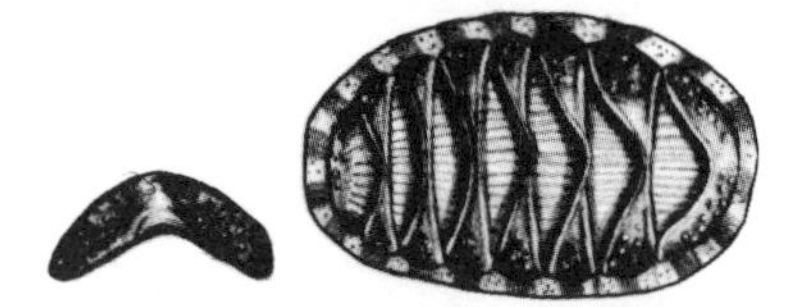

ATLANTIC CHITON. *Chaëto-pleura apiculata*—like the limpet, the chiton has a domelike shell to aid water runoff and a strong foot for holding to the substrate. The chiton also is a rock scraper, grazing with its radula on the surface film of algae.

SPINY BRITTLE STAR. *Ophiothrix angulata*—is a tiny delicate relative of the starfish and hides among plants and sessile animals on the jetty and piling, particularly in calm water areas. The brittle star feeds mainly on the scant detritus caught in crevices in the substrate and supplements this food with larger bits of dead animals. Tiny tube feet seek out food particles and pass them to the mouth on the undersurface. Another brittle star, *Ophioderma brevispinum*, also common to the community, is somewhat larger and dark green, with longer slender arms.

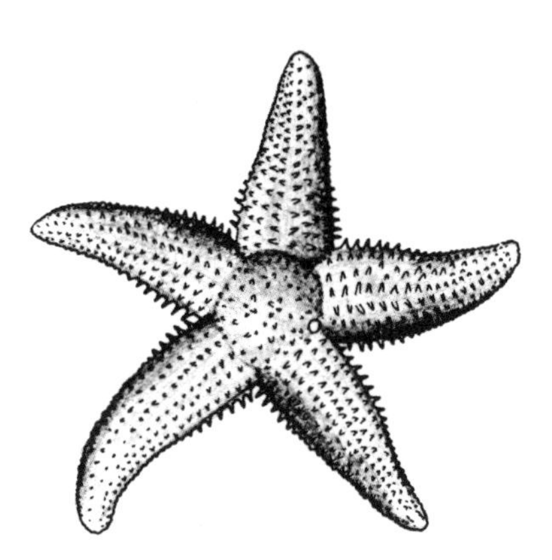

ATLANTIC STARFISH. *Asterias forbesi*—advances up and down the jetty and piling with the tides, preying on molluscs such as clams, mussels, scallops, and oysters. Hundreds of small sensitive tube feet on the undersurface of each arm hold to the substrate and allow the starfish to search rapidly for food. Feeding is accomplished by attaching the tube feet to a bivalve and pulling the valves apart. The starfish then extrudes its own stomach into the bivalve to envelop the victim's soft tissues, and secretes digestive juices which "dissolve" the prey's flesh.

Supratidal life

At the uppermost point of the intertidal zone and at times extending slightly into the splash zone, tiny high tide barnacles crowd together in a thin, horizontal line. In the supratidal and splash zones isopods reign supreme.

HIGH TIDE BARNACLE. *Chthamalus fragilis*—is small, fragile, and rather flattened. The nonporous plates of the shell help prevent water loss during prolonged periods of exposure to air. Like other barnacles, this one filters plankton, obtaining sufficient amounts to survive even in the splash zone.

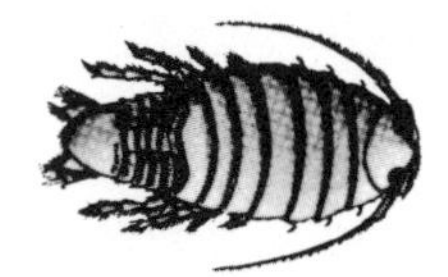

SEA COCKROACH. *Lygida* sp.—is an isopod, up to 2.5 cm long, which looks like its namesake. It scurries over the substrate or hides in crevices. Although the sea cockroach darts in and out of the water briefly, it will drown if submerged for long. A scavenger, it feeds on debris and detritus caught on the substrate.

Additional Resource Books

Amos, William. *The Life Of The Seashore*. New York: McGraw-Hill Book Company, 1966.
Arnold, Augusta Foote. *The Sea-Beach At Ebb-Tide*. New York: Dover Publications, Inc., 1968.
Barnes, Robert D. *Invertebrate Zoology*. 3rd edition. Philadelphia: W.B. Saunders Company, 1974.
Buck, Margaret Waring. *Along The Seashore*. New York: Abingdon Press, 1964.
Carson, Rachel. *The Edge of the Sea*. Boston: Houghton Mifflin Co., 1955.
Clark, John (Editor). "Barrier Islands and Beaches." *Technical Proceedings of the 1976 Barrier Islands Workshop*. Washington: The Conservation Foundation, 1976.
Crowder, William. *Seashore Life Between The Tides*. New York: Dover Publications, Inc., 1959.
Godfrey, Paul J. and Melinda M. Godfrey. "Barrier Island Ecology of Cape Lookout National Seashore and Vicinity, North Carolina." *National Park Service Monograph Series, No. 9*. Washington: U.S. Government Printing Office, 1976.
Gosner, Kenneth L. *Guide to Identification of Marine and Estuarine Invertebrates. Cape Hatteras to the Bay of Fundy*. New York: Wiley-Interscience, 1971.
Gosner, Kenneth L. *A Field Guide to the Atlantic Seashore, From the Bay of Fundy to Cape Hatteras*. Boston: Houghton Mifflin Co., 1979.
Graetz, Karl E. *Seacoast Plants Of The Carolinas*. Raleigh, N.C.: U.S. Department of Agriculture, Soil Conservation Service, 1973.
MacGinitie, G.E. and Nettie MacGinitie. *Natural History of Marine Animals*. New York: McGraw-Hill Book Company, Inc., 1949.
McConnaughey, Bayard H. *Introduction to Marine Biology*. Saint Louis: The C.V. Mosby Company, 1974.
Miner, Roy Waldo. *Field Book of Seashore Life*. New York: G.P. Putnam's Sons, 1950.
Moore, Hilary B. *Marine Ecology*. New York: Wiley Publishing Company, 1958.
Odum, Eugene P. *Fundamentals of Ecology*. Philadelphia: W.B. Saunders Company, 1959.
Perkins, E.J. *The Biology of Estuaries and Coastal Waters*. New York: Academic Press, 1974.
Perlmutter, Alfred. *Guide To Marine Fishes*. New York: Branhall, 1961.
Peterson, Roger Tory. *A Field Guide To The Birds*. Boston: Houghton Mifflin Company, 1947.
Porter, Hugh and Jim Tyler. *Sea Shells Common To North Carolina*. Raleigh, N.C.: N.C. Department of Natural and Economic Resources, Division of Marine Fisheries, 1971.
Radford, Albert E., Harry E. Ahles and C. Ritchie Bell. *Manual Of The Vascular Flora Of The Carolinas*. Chapel Hill: The University of North Carolina Press, 1964.
Ranwell, D.S. *Ecology of Salt Marshes and Sand Dunes*. London: Chapman and Hall, 1975.
Robbins, Chandler S., Bertel Bruun and Herbert S. Zim. *A Guide To Field Identification, Birds of North America*. New York: Golden Press, 1966.
Southward, Alan J. *Life on the Sea-shore*. Cambridge, Mass.: Harvard University Press, 1967.
Van Dover, Cindy and William W. Kirby-Smith. *Field Guide to Common Marine Invertebrates of Beaufort, N.C.* Beaufort, N.C.: Duke University Marine Laboratory, 1979.
Williams, Austin B. "Marine Decapod Crustaceans of the Carolinas." *Fishery Bulletin*, Vol. 65, No. 1, pp. 1-298. Washington: U.S. Department of the Interior, U.S. Fish and Wildlife Service, 1965.
Zim, Herbert S. and Lester Ingle. *Seashores*. New York: Golden Press, 1955.

GLOSSARY

(Terms are defined in the special context of this guide)

Adapted: Able to live in a particular environment.
Aeration: The process of exposing soil or water to air, which increases the oxygen content.
Air bladders: Tiny balloonlike, gas filled structures; aid certain algae in keeping afloat.
Amphipod: A member of the order Amphipoda in the arthropod class Crustacea, with a body flattened from side to side and legs used for both walking and swimming.
Anaerobic: Absence of oxygen; the ability of an organism to carry out its metabolic processes without oxygen.
Annelid: A member of the invertebrate phylum Annelida, with a wormlike, segmented (ringed) body; includes earthworms, leeches, lugworms, and others.
Antennae: Paired appendages which are sensory feelers on the anterior end (head) of arthropods.
Arachnid: A member of the class Arachnida of the phylum Arthropoda; includes horseshoe crabs, spiders, scorpions, and ticks.
Arthropod: A member of the invertebrate phylum Arthropoda, which includes insects, crustaceans, arachnids, and other animals with jointed appendages.

Barrier islands: The islands which parallel the coast of North Carolina and act as barriers for protecting the mainland from the full effects of ocean forces.
Bloom: Rapid and unusually large seasonal increase in certain planktonic algae and other minute organisms, either in the ocean or in sounds.
Brackish: Moderately saline water; seawater that has been significantly diluted with fresh water; see also *salinity*.
Bryozoan: A member of the invertebrate phylum Bryozoa (or Ectoprocta), occurring either as attached mosslike colonies or as flat, encrusting colonies on solid objects.
Byssal threads: Fibers or tufts of tiny filaments produced by certain molluscs to attach themselves securely to an object.

Calcareous: Containing calcium carbonate which hardens into a solid, inflexible substance.
Carapace: A tough, horny outer covering, often composed of chitin, which provides a protective shield for crustaceans and some other arthropods.
Carnivore: An animal that eats only other animals to obtain nutrients and energy.
Chemosensitivity: The ability to sense chemicals in the environment.
Chordate: A member of the vertebrate phylum Chordata which includes tunicates, birds, fishes, mammals, and other animals that have a notochord and a tubular nerve cord.
Cilia: Tiny, hairlike projections, generally capable of motion, which may aid in sensing, gathering food, and moving.
Cirri: Tiny, hairlike appendages, used primarily as sensing devices, often fused cilia.
Coelenterate: A member of the invertebrate phylum Coelenterata which includes jellyfish, corals, hydroids, and other animals having a multiple-use body cavity with a single opening.
Colonial: Describes members of the same species living as a group in close association; not necessarily dependent upon one another for survival, but often bound together by an exterior skeletal structure.
Community: A group of plant and animal species living together in a habitat and having unique relationships, particularly those of a food web; ecologists view communities as interdependent units of life in nature, each exhibiting an organization and a structure.
Consumer: An animal which depends upon the organic tissues of other organisms for nutrients and energy.
Copepod: A member of the order Copepoda in the arthropod class Crustacea; small and aquatic with a rounded body and oar-shaped swimming appendages.

Crustacean: A member of the class Crustacea of the invertebrate phylum Arthropoda; includes shrimps, crabs, barnacles, copepods and other animals having jointed legs, segmented bodies, and a hard external skeleton.

Ctenophore: A member of the invertebrate phylum Ctenophora having an oval, transparent jellylike body with eight rows of comblike plates.

Debris: Pieces and particles of decaying plant and animal matter larger than detritus.

Decomposer: An organism like a bacterium or fungus which uses tissues of dead plants and animals as an energy and nutrient source.

Decomposition: The breakdown of plant and animal remains from organic to inorganic matter, accomplished primarily through the metabolic and other processes of bacteria and fungi, but also through other environmental factors such as wind, heat, water, and waves.

Desiccation: Severe loss of water, often involving dehydration of the body, due largely to evaporation.

Detritus: Minute particles of decaying organic matter.

Deposit feeder: An animal that feeds on materials deposited on a substrate.

Diurnal: Active during the day.

Drift line: See "strand line".

Echinoderm: A member of the invertebrate phylum Echinodermata, including starfish, sand dollars, sea urchins, and other animals having a water vascular system and a hard, radially symmetrical, spiny body.

Ecology: The study of all the relationships between organisms and their environment.

Environment: All the conditions, factors, and influences, living and nonliving, which surround an organism or group of organisms.

Estuaries: Water areas where salt water and fresh water mix to produce intermediate salinities, such as in the coastal sounds and river mouths of North Carolina.

Feeding types: Categories of feeding habits among animals, which describe how and where an organism obtains its food.

Filter feeder: An animal which strains or filters water flowing through or around its body to capture suspended food particles.

Flagellae: Tiny whiplike projections, longer than cilia, used in locomotion or to create a current of water.

Food finder: A slow moving animal which scavenges for food, or one which "grazes" for food by scraping the surface.

Food web: All the interrelated food chains or feeding relationships in a community, including producers, consumers, and decomposers, by which energy and nutrients are passed through a community.

Frond: A leafy or leaflike part of a plant, including seaweed.

Gastropod: a member of the class Gastropoda of the invertebrate phylum Mollusca; includes snail-type animals and others which have a one piece shell or no shell, and a broad ventral "foot".

Gelatinous: Jellylike, as the viscous substance coating the outer layer of some plants and animals to prevent desiccation.

Genus: A classification category of plants or animals with common distinguishing characteristics; may consist of many or single species; the first part of an organism's scientific name, capitalized and followed by the species name which generally is not capitalized.

Grazer: An animal which moves slowly over a surface, feeding on organisms found there.

Habitat: The location where an organism lives, including its surrounding environment.

Hemichordate: A member of the animal phylum Hemichordata; a wormlike marine organism.
Herbivore: An animal that eats only plants to obtain its nutrients and energy.
Holdfast: A specialized body structure that some organisms use to attach to the substrate or other support.
Host: An organism on or in which another organism lives, as the host in a commensal relationship where the host is unharmed and the commensal generally benefits from the association.

Inorganic: Refers to compounds that seldom contain carbon and are not of animal or plant origin; they are essential in living processes and are obtained by plants primarily from soil and water.
Interstitial life: Also called psammon; microscopic forms of life that exist on and between the grains of sand substrate, and which are no bigger than the grains.
Intertidal: That zone of the shoreline habitat between the low and high tide lines, alternately exposed to air and covered by water during the daily tidal cycle.
Invertebrate: An animal without an internal spinal column or "backbone"; all animals other than fishes, amphibians, reptiles, birds, and mammals.
Isopod: A member of the order Isopoda in the class Crustacea, phylum Arthropoda, with a small body, flattened from top to bottom, and many legs generally equal in size and form.

Marine: Of or relating to the oceans and salt water.
Marsh bank: A sharp drop in elevation at the low tide line; this change in topography results from the accumulation of silt and mud caught and retained by marsh grass over a long period of time.
Metabolic processes (metabolism): The chemical and physical processes occurring in the bodies and cells of organisms.
Motile: Having the ability to move without dependence on external forces.
Mucous net: A mucous or slime coating which traps organic food particles filtered from water circulating over gill filaments; characteristic of bivalve molluscs and some worms.

Neap tide: Daily high and low tides which occur in association with the second and fourth quarter phases of the moon; characterized by tides which are not as low or as high as those of spring tides (see spring tides).
Nematocysts: Minute stinging structures in some tentacle cells of certain invertebrates, notably coelenterates, some of which contain poison for paralyzing prey.
Niche: The role a species plays in the community; with respect to feeding type, each species has its own particular niche as producer, consumer, or decomposer.
Nocturnal: Active at night.
Nodes: Swellings; the part of a plant stem from which branches, leaves, and flowers develop.

Omnivore: An animal that eats both plants and other animals to obtain nutrients and energy.
Operculum: A lidlike covering; the hardened flap or horny plate in gastropods which serves as a protective "door," sealing the opening to the shell.
Organic: Substances or compounds containing carbon, derived from plants and animals; they are a source of energy and nutrients, especially for consumers.
Oxidation: The process by which oxygen combines with another substance, a reaction which releases energy.

Parapodia: Paired lobes, containing many setae and cilia, on the side of each body segment of marine worms (polychaetes).
Pelecypod: A member of the class Pelecypoda of the invertebrate phylum Mollusca, with a body contained in a two-part (bivalve) shell; clams, oysters, mussels, etc.

Phytoplankton: Floating microscopic plants which are important as producers in marine food webs.
Pioneer species: Species which are the first to settle in an uninhabited area; very hardy and able to withstand a harsh environment.
Plankton: Usually microscopic plants and animals floating in the sea; constitute the first and second steps in marine food webs.
Planktonic: Swept about without will, unable to move against the effects of tides and currents; such an existence is the result of small size or ineffective structures for self movement (motility).
Polyp: Sessile stage in the life history of all coelenterates; has a mouth surrounded by small, slender tentacles used as sensing devices to capture plankton and other food.
Proboscis: A tubelike organ, usually part of the mouth, which in some animals can be extended and retracted to aid in sensing, sucking, and capturing food.
Producer: A green plant, able through photosynthesis to produce organic material as part of its body tissues; producers compose the first step in a community food web.
Production (productive): The amount or rate of organic material produced by plants during photosynthesis.
Psammon: See "interstitial life".
Pycnogonid: A member of the class Pycnogonida of the invertebrate phylum Arthropoda; has a small body and long legs.

Radula: A straplike device, covered with rows of small "teeth", that is part of the mouth structure in certain molluscs; used for scraping, tearing, boring, and ingesting food.
Rakers: Fingerlike projections on the gills of fishes; rakers clean and clear the surfaces of gill filaments.
Red tides: Large increases or blooms in the numbers of certain microscopic organisms in the sea, notably flagellates, which cause the water to appear red.
Rostrum: A pointed extension of the carapace in the head region of crustaceans; acts as a protective device.

Salinity: A measurement of dissolved salts; ocean waters generally average 35 0/00 (35 parts of dissolved salts per 1000 parts of water); intermediate salinities (15 0/00 to 30 0/00) occur in coastal sounds; lower salinities (1.0 0/00 to 10 0/00) occur in brackish waters where coastal rivers flow into sounds.
Salt flats: Relatively barren areas, more elevated than the surrounding topography, in the intertidal marsh; evaporation in the salt flats leaves high salt residues that prevent the survival of all but the most salt tolerant plant species.
Scraper: An animal which scrapes its food from some surface.
Sediments: Particles which settle to the bottom or which compose the bottom material (substrate) on which an organism lives.
Sediment ingester: An animal which eats quantities of sediments, extracting organic matter and then defecating the remaining material.
Sessile: Describes a plant or animal which attaches itself permanently to an object or surface.
Shoal: A shallow area in the water, such as a sand bar elevated above the surrounding bottom topography.
Siphon: A tubelike organ in some marine animals, used for moving currents of water into and out of the body to extract oxygen and planktonic food and to remove wastes; an incurrent siphon draws in water and an excurrent siphon ejects water.
Sounds: Areas of water which on the North Carolina coast separate the mainland from the outer barrier islands.
sp.: Abbreviation of the word species; used where the species name cannot be given for any of a variety of reasons.

Species: A basic classification category of plants or animals which describes and divides organisms of a single genus into those which share the highest degree of similarity and which can breed with one another to produce offspring capable of living and reproducing; the second part of an organism's scientific name, which follows the genus (generic) name and is not capitalized.

Splash zone: That area of the shoreline, particularly on jetties and pilings, where waves striking the habitat send a splash or spray of water into what otherwise is the supratidal zone.

Spring tides: Daily high and low tides which occur in association with the first and third (new and full moon) phases of the moon; characterized by the highest and lowest tides of the month.

Strand line: That part of the shoreline where debris and detritus deposited by the high tide become stranded.

Substrate: The bottom material on or in which an organism lives or to which it is attached.

Subtidal: Refers to the zone of the shoreline habitat below the low tide line, always covered by water.

Supratidal: Refers to the zone of the habitat above the high tide line where tidal water generally does not reach.

Suspension feeder: An animal that filters or screens suspended food particles from the water.

Swarm: The occurrence of unusually large numbers of certain animals, generally seasonal.

Swimmerettes: Appendages on the abdomen of many kinds of crustaceans, usually like tiny legs or paddles, which aid in movement, in creating currents of water, and in carrying eggs.

Tentacles: Long, slender, flexible projections in the head or mouth region of some invertebrates, like coelenterates and certain worms; used for grasping, sensing, and feeding.

Tidal cycle: The cycle of high and low tides occurring twice during a 24-hour period on the coast of North Carolina.

Topography: The elevations or contours (surface features) of the substrate.

Transition organism: An organism or species believed to be in transition from one medium to another in its living habits (air to water habitats or vice versa); generally a longterm process.

Tube feet: Projections occurring in large numbers on the bodies of most echinoderm species; each tube foot ends in a suction disc, and they collectively aid in movement, in holding to the substrate, and in opening the shells of bivalve molluscs.

Vertebrate: A member of the subphylum Vertebrata of the phylum Chordata; an animal with a segmented spinal column or "backbone" containing a central spinal cord; fishes, amphibians, reptiles, birds, and mammals.

Zonation: Distribution of plants and animals into zones or specific areas which differ from each other in species composition; caused by different environmental conditions in each zone; along the shoreline, zones are often parallel horizontal or vertical bands, maintained by tidal levels.

Zooplankton: Floating microscopic animals, including the larval stages of many larger animals; feed on phytoplankton and other zooplankton and, in turn, are an important food source for larger animals in marine food webs.

INDEX

Scientific Names

Common Names